ADVANCES IN KINETICS AND MECHANISM OF CHEMICAL REACTIONS

ADVANCES IN KINETICS AND MECHANISM OF CHEMICAL REACTIONS

Edited by

Gennady E. Zaikov, DSc, Artur J. M. Valente, PhD, and Alexei L. Iordanskii, DSc

Apple Academic Press

TORONTO NEW JERSEY

Apple Academic Press Inc.
3333 Mistwell Crescent
Oakville, ON L6L 0A2
Canada

Apple Academic Press Inc.
9 Spinnaker Way
Waretown, NJ 08758
USA

Printed in the United States of America on acid-free paper

International Standard Book Number-13: 978-1-926895-42-0 (Hardcover)

Library of Congress Control Number: 2012951941

Library and Archives Canada Cataloguing in Publication

Advances in kinetics and mechanism of chemical reactions/edited by Gennady E. Zaikov, Artur J.M. Valente, and Lexei L. Iordanskii.

Includes bibliographical references and index.
ISBN 978-1-926895-42-0
1. Chemical kinetics. 2. Chemical systems. 3. Reactivity (Chemistry). I. Zaikov, G. E. (Gennadiˇi Efremovich), 1935- II. Valente, Artur J. M III. Iordanskiˇi, Lexei L

QD502.A39 2013 541'.394 C2012-906388-6

Apple Academic Press also publishes its books in a variety of electronic formats. Some content that appears in print may not be available in electronic format. For information about Apple Academic Press products, visit our website at **www.appleacademicpress.com** and the CRC Press website at **www.crcpress.com**

About the Editors

Gennady E. Zaikov, DSc

Gennady E. Zaikov, DSc, is Head of the Polymer Division at the N. M. Emanuel Institute of Biochemical Physics, Russian Academy of Sciences, Moscow, Russia, and Professor at Moscow State Academy of Fine Chemical Technology, Russia, as well as Professor at Kazan National Research Technological University, Kazan, Russia. He is also a prolific author, researcher, and lecturer. He has received several awards for his work, including the the Russian Federation Scholarship for Outstanding Scientists. He has been a member of many professional organizations and is on the editorial boards of many international science journals.

Artur J. M. Valente, PhD

Artur J. M. Valente received his PhD from Coimbra University, Portugal, in 1999 working on the transport properties of non-associated electrolytes in hydrogels. He is currently Assistant Professor in the Chemistry Department of University of Coimbra. In 2004 and 2006 he was a guest and invited researcher, respectively, in the Division of Physical Chemistry 1, Lund University, Sweden. His research interests focus on the transport properties of ionic and non-ionic solutes in multicomponent systems, such as host-guest compounds, as well as in the characterization of the transport properties in polymeric matrices, with particular emphasis to polyelectrolytes, gels and functional blends, and composites. He has more than 90 publications in ISI international journals, over 100 communications at scientific meetings, 10 papers in national journals, one patent, and two books. He is also co-editor of two books, a special issue of the Journal of Molecular Liquids (2010), and a forthcoming issue of *Pure & Applied Chemistry*.

Alexei L. Iordanskii, DSc

Alexei L. Iordanskii, DSc, is Professor at the Institute of Chemical Physics at the Russian Academy of Sciences in Moscow, Russia. He is a scientist in the field of chemistry and the physics of oligomers, polymers, composites, and nanocomposites.

Contents

List of Contributors

L. A. Badykova
Institute of Organic Chemistry, Ufa Scientific Center, Russian Academy of Sciences,
pr. Oktyabrya 69, Ufa, 450054, Bashkortostan, Russia
E-mail: badykova@mail.ru

D. V. Bagrov
Faculty of Biology, Moscow State University, Leninskie gory 1-12, 119992 Moscow, Russia

K. Bauerová
Institute of Experimental Pharmacology and Toxicology, Slovak Academy of Sciences, SK-84104, Bratislava, Slovakia

L. I. Bazylyak
Chemistry of Oxidizing Processes Division; Physical Chemistry of Combustible Minerals Department; Institute of Physical–organic Chemistry & Coal Chemistry named after L. M. Lytvynenko; National Academy of Science of Ukraine
3a Naukova Str., Lviv–53, 79053, UKRAINE; E–mail: hav.ok@yandex.ru

A. P. Bonartsev
A. N. Bach's Institute of Biochemistry, Russian Academy of Sciences, Leninskiy prosp. 33, 119071 Moscow, Russia
Faculty of Biology, Moscow State University, Leninskie gory 1-12, 119992 Moscow, Russia

G. A. Bonartseva
A. N. Bach's Institute of Biochemistry, Russian Academy of Sciences, Leninskiy prosp. 33, 119071 Moscow, Russia

A. P. Boskhomodgiev
A. N. Bach's Institute of Biochemistry, Russian Academy of Sciences, Leninskiy prosp. 33, 119071 Moscow, Russia

F. Dráfi
Institute of Experimental Pharmacology and Toxicology, Slovak Academy of Sciences, SK-84104, Bratislava, Slovakia

E. A. Filatova
A. N. Bach's Institute of Biochemistry, Russian Academy of Sciences, Leninskiy prosp. 33, 119071 Moscow, Russia

A. K. Haghi
University of Guilan, Rasht, Iran
E-mail: Haghi@Guilan.ac.ir

A. L. Iordanskii
A. N. Bach's Institute of Biochemistry, Russian Academy of Sciences, Leninskiy prosp. 33, 119071 Moscow, Russia
N. N. Semenov Institute of Chemical Physics, Russian Academy of Sciences, Kosygin str. 4, 119991 Moscow, Russia
E-mail: aljordan08@gmail.com, tel. +74959397434, fax: +74959382956

E. A. Ivanov
A. N. Bach's Institute of Biochemistry, Russian Academy of Sciences, Leninskiy prosp. 33, 119071 Moscow, Russia

O. Yu. Khavunko
Chemistry of Oxidizing Processes Division; Physical Chemistry of Combustible Minerals Department; Institute of Physical–organic Chemistry & Coal Chemistry named after L. M. Lytvynenko; National Academy of Science of Ukraine
3a Naukova Str., Lviv–53, 79053, UKRAINE; E–mail: hav.ok@yandex.ru

E. I. Korotkova
Tomsk Polytechnic University, 30 Lenin Street, 634050, Tomsk, Russia

G. V. Kozlov
Institute of Applied Mechanics of Russian Academy of Sciences,
Leninskii pr., 32 a, Moscow 119991, Russian Federation

Victor M. M. Lobo
Departamento de Química, Universidade de Coimbra,
3049 Coimbra, Portugal
E-mail: anacfrib@ci.uc.pt

T. K. Makhina
A. N. Bach's Institute of Biochemistry, Russian Academy of Sciences, Leninskiy prosp. 33, 119071 Moscow, Russia

V. S. Maltseva
Candidate of Chemical Sciences, Docent of the Department "General and Inorganic Chemistry", Southwest State University

Yu. G. Medvedevskikh
Chemistry of Oxidizing Processes Division; Physical Chemistry of Combustible Minerals Department; Institute of Physical–organic Chemistry & Coal Chemistry named after L. M. Lytvynenko; National Academy of Science of Ukraine
3a Naukova Str., Lviv–53, 79053, UKRAINE; E–mail: hav.ok@yandex.ru

V. M. Misin
Emanuel Institute of Biochemical Physics Russian Academy of Sciences, 4 Kosygin Street, 119334, Moscow, Russia, E-mail: Natnik48s@yandex.ru

D. Mislovičová
Institute of Chemistry, Slovak Academy of Sciences, SK-84538 Bratislava, Slovakia

Yu. B. Monakov
Institute of Organic Chemistry, Ufa Scientific Center, Russian Academy of Sciences,
pr. Oktyabrya 69, Ufa, 450054,Bashkortostan, Russia
E-mail: badykova@mail.ru

R. Kh. Mudarisova
Institute of Organic Chemistry, Ufa Scientific Center, Russian Academy of Sciences,
pr. Oktyabrya 69, Ufa, 450054,Bashkortostan, Russia
E-mail: badykova@mail.ru

V. L. Myshkina
A. N. Bach's Institute of Biochemistry, Russian Academy of Sciences, Leninskiy prosp. 33, 119071 Moscow, Russia

M. Nagy
Department of Pharmacognosy and Botany, Faculty of Pharmacy, Comenius University, SK-83232, Bratislava, Slovakia

Joaquim J. S. Natividade
Departamento de Química, Universidade de Coimbra,
3049 Coimbra, Portugal
E-mail: anacfrib@ci.uc.pt

F. F. Niyazi
Doctor of Chemical Sciences, Professor, Head of the Department "General and Inorganic Chemistry", Southwest State University

E. Priesolová
Department of Pharmacognosy and Botany, Faculty of Pharmacy, Comenius University, SK-83232, Bratislava, Slovakia

Yu. I. Puzin
Ufa State Petroleum Technological University, 1 Kosmonavtov Street,
450062, Ufa, Russian Federation
E-mail: ppuziny@rambler.ru

P. Rapta
Slovak University of Technology, Faculty of Chemical and Food Technology, Institute of Physical Chemistry and Chemical Physics, SK-81237, Bratislava, Slovakia

A. V. Rebrov
A. V. Topchiev Institute of Petroleum Chemistry. Leninskiy prosp. 27, 119071 Moscow, Russia

Ana. C. F. Ribeiro
Departamento de Química, Universidade de Coimbra,
3049 Coimbra, Portugal
E-mail: anacfrib@ci.uc.pt

N. N. Sazhina
Emanuel Institute of Biochemical Physics Russian Academy of Sciences, 4 Kosygin Street, 119334, Moscow, Russia, E-mail: Natnik48s@yandex.ru

A. V. Sazonova
Postgraduate student of the Department "General and Inorganic Chemistry", Southwest State University

M. Slováková
Slovak University of Technology, Faculty of Chemical and Food Technology, Institute of Physical Chemistry and Chemical Physics, SK-81237, Bratislava, Slovakia

L. Šoltés
Institute of Experimental Pharmacology and Toxicology, Slovak Academy of Sciences, SK-84104, Bratislava, Slovakia

K. Valachová
Institute of Experimental Pharmacology and Toxicology, Slovak Academy of Sciences, SK-84104, Bratislava, Slovakia

Artur J. M. Valente
Departamento de Química, Universidade de Coimbra, 3049 Coimbra, Portugal
E-mail: anacfrib@ci.uc.pt

S. A. Yakovlev
A. N. Bach's Institute of Biochemistry, Russian Academy of Sciences, Leninskiy prosp. 33, 119071 Moscow, Russia

Yu. G. Yanovskii
N. M. Emanuel Institute of Biochemical Physics
of Russian Academy of Sciences, Kosygin st., 4, Moscow 119334, Russian Federation

E. I. Yarmukhamedova
Institute of Organic Chemistry of Ufa' Research Center of Russian Academy of Sciences, 71 Octyabrya Blvd. 450054, Ufa, Russian Federation
E-mail: gluhov_e@anrb.ru

G. E. Zaikov
Chemistry of Oxidizing Processes Division; Physical Chemistry of Combustible Minerals Department; Institute of Physical–organic Chemistry & Coal Chemistry named after L. M. Lytvynenko; National Academy of Science of Ukraine
3a Naukova Str., Lviv–53, 79053, UKRAINE; E–mail: hav.ok@yandex.ru

G. E. Zaikov
N. M. Emanuel Institute of Biochemical Physics of Russian Academy of Sciences, Kosygin st., 4, Moscow 119334, Russian Federation Russian Academy of Sciences, Moscow, Russia
E_mail: GEZaikov@Yahoo.com

List of Abbreviations

ABTS	2,2'-Azino-bis(3-ethylbenzothiazoline-6-sulphonic acid)
AFM	Atomic force microscopy
AG	Arabinogalactan
AIBN	Azobisisobutyronitrile
AO	Antioxidants
BAS	Bioanalytical system
BAS	Biological active substances
BP	Benzoyl peroxide
BSR	Butadiene-styrene rubber
CV	Cyclic voltammetry
DC	Diketocarboxylic acids
DFM	Dynamic force microscope
DHA	Dehydroascorbate
DMPO	5,5-Dimethyl-1-pyrroline-N-oxide
DMSO	Dimethyl sulfoxide
DST	Department of Science and Technology
EHD	Electrohydrodynamic
EMI	Electromagnetic interference
EMISE	Electromagnetic interference shielding effectiveness
EPR	Electron paramagnetic resonance
FTIR	Fourier transforms infrared
GA	Gallic acid
GAGs	Glycosaminoglycans
GC	Glassy carbon
HA	Hyaluronan
HH	3-Hydroxyheptanoate
HO	3-Hydroxyoctanoate
HV	3-Hydroxyvalerate
INAH	Isonicotinic acid hydrazide
IR	Infrared
LBM	Lattice boltzmann method
MFE	Mercury film electrode
MMA	Methyl methacrylate
MW	Molecular weight
MWD	Molecular weight distribution
NDSA	Naphtalene disulfonic acid
NMP	N-methyl-2 pyrolidon
NMR	Nuclear magnetic resonance
NR	Natural rubbers
ODE	Ordinary differential equation
PAN	Polyacrylonitrile

PANI	Polyaniline
PANIEB	Polyaniline emeraldine base
PB	Polybutadiene
PHAs	Polyhydroxyalkanoates
PHB	Poly(3-R-hydroxybutyrate)
PHBV	Poly(3-hydroxybutyrate-co-3-hydroxyvalerate)
PLA	Poly(L-lactide)
PMMA	Poly(methyl methacrylate)
PMR	Proton magnetic resonance
PPy	Polypyrrole
PVA	Polyvinyl acetate
RMSD	Root-mean-square deviation
ROS	Reactive oxygen species
ROX	Redox initiating systems
SAWS	Self-avoiding walks statistics
SE	Shielding effectiveness
SEC	Size exclusion chromatography
SEM	Scanning electron microscope
SF	Synovial fluid
SPIP	Scanning probe image processor
SPM	Scanning probe microscopy
STM	Scanning tunneling microscopy
TBAP	Tetrabutylammonium perchlorate
TC	Technical carbon
TEAC	Trolox equivalent of antioxidant capacity
TEM	Transmission electron microscopes
THF	Tetrahydrofuran
THP	Theophylline
TMS	Tetramethylsilane
TMT	1,3,5-trimethyl-hexahydro-1,3,5-triazine
UCM	Upper-convected maxwell
WAXS	Wide angle X-ray scattering
WDX	Wavelength-dispersive X-ray
XRD	X-ray diffraction Technique

Introduction

Advances in Kinetics and Mechanism of Chemical Reactions describes the chemical physics and/or chemistry of 10 novel material or chemical systems. These 10 novel material or chemical systems are examined in the context of issues of structure amd bonding, and/or reactivity, and/or transport properties, and/or polymer properties, and/or biological characteristics. This eclectic survey thus encompasses a special focus on the associated kinetics, reaction mechanisms and/or other chemical physics properties, of these 10 broadly chosen material or chemical systems. Thus, the most contemporary chemical physics methods and principles are applied to the characterization of the properties of these 10 novel material or chemical systems. The coverage of these novel systems is thus broad, ranging from the study of biopolymers to the analysis of antioxidant and medicinal chemical activity, on the one hand, to the determination of the chemical kinetics of novel chemical systems, and the characterization of elastic properties of novel nanometer scale material systems, on the other hand.

Advances in Kinetics and Mechanism of Chemical Reactions is divided into 10 chapters.

Chapter 1, by Valachova et al., describes their chemical system as comprised of the compound "arbutin" in cupric ion solution with ascorbate as a reagent for treating hyaluronan (HA). Thusarbutin was tested in the function of a potential anti- or prooxidant in Cu(II) plusascorbate, and it induced degradation of high-molar-mass hyaluronan (HA). The time- and dose-dependences of dynamic viscosity changes of the HA solutions were investigated by the method of rotational viscometry. Both the reduction of the dynamic viscosity of the HA solution and the decrease of the polymer mean molar mass as revealed by the method of size exclusion chromatography proved the tenet that on using the Cu(II) ions plusascorbate, i.e. the Weissberger's oxidative system, the degradation of HA macromolecules is pronounced by added arbutin. These studies of hyaluronan (HA), a biopolymer consisting of disaccharide units, are important from the perspective of understanding the physiology of HA in the bones and joints and in other tissues. The medical importance of understanding HA turnover in the human body cannot be underestimated.

In Chapter 2, Ribiero et al., describe their investigations of diffusion kinetics, including the modeling of electrolytes and non-electrolytes. Thus in the past few years, their diffusion phenomena group has been particularly dedicated to the study of mutual diffusion behavior of binary, ternary and quaternary solutions, involving electrolytes and non-electrolytes, helping to go deeply into the understanding of their structure, and aiming at practical applications in fields as diverse as corrosion studies occurring in biological systems or therapeutic uses. In fact, the scarcity of diffusion coefficients and other transport data in the scientific literature, due to the difficulty of their accurate experimental measurement and impracticability of their determination by theoretical procedures, coupled to their industrial and research need, well justify the work reported here by Ribiero et al. in accurate measurements of such transport properties.

Chapter 3 describes the frictional and elastic properties of the chemical system comprised of polystyrene dissolved in toluene. Medvedevskikh et al. describe their work thus, as it has been experimentally investigated by the gradient dependence of the effective viscosity η for concentrated solutions of polystyrene in toluene at three concentrations, ρ, and for the four fractions of polystyrene characterized by four distinct molar weights, M. The gradient dependence of each respective solution's viscosity was studied at four temperatures, T, for each pair of ρ and M valuations. The experiments were carried out with the use of a standard viscosity meter at the different angular velocities ω (turns/s) of the working cylinder rotation. An analysis of the η(ω) dependencies permitted the marking out of the frictional (ηf) and elastic (ηe) components of the viscosity and to study their dependence on temperature T, concentration ρ and the length of a chain N. These fundamental chemical physics studies, carried out on the molecular scale, are important for the theory underlying the viscosity phenomena. Where viscosity is a physical quantity that certainly requires further investigations and measurements on the molecular scale.

In Chapter 4 of this volume concerning reinforcement mechanisms of nanocomposites, Kozlov et al. describe the theoretical and experimental study of the mechanics of nanoscopic matter. The modern methods of experimental and theoretical analysis of polymer materials structure and properties have allowed Kozlov et al. to confirm earlier propounded hypotheses, but also to obtain principally new results. These investigators consider some important problems of particulate-filled polymer nanocomposites, the solution of which allows one to advance substantially the understanding of these materials and their unusual properties. It thus endows one with the ability to understand and predict. In this aspect interfacial regions play a particular role, since it has been shown earlier that they are the same reinforcing element in elastomeric nanocomposites, and they thus occur as nanofiller actually. Therefore the knowledge of interfacial layer dimensional characteristics is necessary for quantitative determination of one of the most important parameters of polymer composites in general – their reinforcement degree. And thus this study by Kozlov et al. is an important investigation of the chemical physics of nanocomposites, including a clarification of their multicomponent nature and a description of the nanocomposite reinforcing elements.

In Chapter 5, Mudarisova et al. describe their work in drug discovery, in the area of antituberculosis preparations. In particular, this group describes the addition of known tuberculostatic drugs to saccharides, in an attempt to develop medicinal compositions that can overcome drug resistance in bacteria causing tuberculosis. Thus the work of Mudarisova et al. can be summarized, drug discovery is one of the thrust areas of modern medicinal chemistry. The search for and development of new antituberculosis agents have recently become of interest because of the drug resistance of mycobacteria to existing drugs. One promising direction for creating such drugs is the addition of common tuberculostatics to polysaccharides. It is known that the polysacccharidearabinogalactan (AG) has a broad spectrum of biological activity. However, its tuberculostatic activity has not been reported. Herein the modification of AG and its oxidized forms by the antituberculosis drug isonicotinic acid hydrazide (INAH) and the antituberculosis activity of the resulting compounds are studied.

Chapter 6 of this volume by Niyazi et al. describe determination of the sorption properties of polymer cellulose and natural carbonate sorbents for use in wastewater treatment applications. Niyazi et al. thus summarize, "the article shows the comparative characteristics of the sorption properties of polymer cellulose and natural carbonate sorbents. The influence of the mass of sorbents on the degree of extraction as well as the pH changing are analyzed. The optimum phase ratio has been determined. Kinetic curves have been plotted." The work of Niyazi et al. thus advances our understanding and development of wastewater treatment reagents for applications of clarifying wastewater in many and varied areas, both industrial and domestic, in everyday life.

Yarmukhamedova et al. in Chapter 7 of this volume investigate chemical physics properties of a class of aromatic compounds (diketocarboxylic acids) on the radical initiation properties of an initiator compound used in a polymerization reaction system. Thus as Yarmukhamedova et al. describe, "the influence of aromatic diketocarboxylic acids on the decomposition initiator of radical polymerization - azobisisobutyronitrile was studied by UV spectroscopy. The interaction occurs with the participation of carboxyl groups of diketocarboxylic acids with nitrile groups of the initiator. It is shown that polymer obtained in the presence of aromatic diketocarboxylic acids has mainly a syndiotactic structure." And thus such work as that reported here by Yarmukhamedova et al. advances our understanding of the synthesis and properties of technologically important classes of radical polymerization polymers.

In Chapter 8 of this volume Yarmukhamedova et al. continue their work on radical polymerization synthesis and the chemical physics properties of the radical initiators used in such synthetic procedures. Yarmukhamedova et al. thus summarize their study of the methyl methacrylate synthetic polymer system, "the influence of the 1,3,5-trimethyl-hexahydro-1,3,5-triazine on the radical polymerization of methyl methacrylate was studied. The kinetic parameters were obtained (reaction orders, activation energy of polymerization). It is established that the triazine is the slight chain transfer agent during to polymerization initiated by azo-bis-isobutyronitrile, and the component of the initiating system if the peroxide initiator is used. Polymers synthesized in the presence of 1,3,5-trimethyl-hexahydro-1,3,5-triazine have the higher content of syndio – and isotactic sequences in the macromolecule." Such fundamental polymer chemical physics work as reported here by Yarmukhamedova et al. advances our understanding of these technologically important classes of polymer systems.

In Chapter 9 of this volume, Bonartsev et al. focus their work on the degradation properties, including particularly the degradation kinetics of the biochemical reagent known aspoly(3-hydroxybutyrate) and its derivatives. This work is designed to be an informative source for research on biodegradable poly(3-hydroxybutyrate) and its derivatives. Bonartsev et al. focus on hydrolytic degradation kinetics at two distinct temperatures, in phosphate buffer to compare polymer kinetic profiles. These investigators report chemical physics properties of these novel biopolymer systems (i.e. poly(3-hydroxybutyrate) due to the economic interest in these natural polymers over the well-known synthetic polymers. For it is well known that natural polymers represent an emerging area of technological interest and application, especially so with their biodegradability

Finally in Chapter 10, Sazhina et al. present analytical measurements of antioxidant content in various food products. The antioxidant measurements are accomplished by means of electrochemical instrumentation. Sazhina et al. thus describe their work, "A comparison of the total content of antioxidants and their activity with respect to oxygen and its radicals in juice and extracts of herbs, extracts of a tea and also in human blood plasma was carried out in the present work by use of two operative electrochemical methods: ammetric and voltammetric. Efficiency of methods has allowed studying dynamics of antioxidants content and activity change in same objects during time. Good correlation between the total phenol antioxidant content in the studied samples and values of the kinetic criterion defining activity with respect to oxygen and its radicals is observed." Thus Sazhina et al. provide an interesting methodology and corresponding analysis of the levels of important antioxidant components within common food products. This work is important for advancing our knowledge of antioxidant biochemistry and its potential therapeutic properties.

In conclusion, we see thus in this volume, **Advances in Kinetics and Mechanism of Chemical Reactions**, that chapters 1, 6 and 9 address the investigation of three novel and distinct biopolymer systems, as to their chemical kinetics and other chemical properties. In Chapter 2 there is then a digression into the measurement of the diffusion kinetics of electrolyte and non-electrolyte systems, with application to corrosion studies. Chapter 3 covers the determination of viscoelastic properties of the chemical system comprised of polystyrene dissolved in toluene. And the reinforcement mechanisms of nanometer scale composites, with special attention focused on mechanical properties of representative nanometer scale composite materials, is reported in Chapter 4. And in Chapter 5, the biological activity of a novel tuberculostatic drug-polysaccharide composition is reported from novel synthetic studies. Then polymer chemical kinetics of two important classes of radical polymerization polymers, with special focus on the radical initiators used in thesepolymer synthetic reactions, provides the focus of Chapters 7 and 8. In this work on polymer chemical physics both reaction orders and activation energies were determined for technologically important classes of radical polymerization polymers. Chapter 10 concludes with an investigation and novel determination of antioxidant content in various foodstuffs by the classical analytical chemistry technique of electrochemistry.

It can be concluded, from this brief survey of the present volume, that broad chemical physics coverage of 10 novel material or chemical systems is reported within its pages. The chemical physics methods used to characterize these 10 novel systems are clearly state-of-the-art, and the results should be intriguing to the prospective readership in chemistry and physics and nanoscience, including those scientists engaged in chemical physics research and the polymer chemistry and physics communities, as well as those researchers involved in biological chemistry research and also those scientists focused on nanotechnology.

— Gennady E. Zaikov

1 Radical Degradation of High Molar Mass Hyaluronan Induced by Ascorbate Plus Cupric Ions Testing of Arbutin in the Function of Antioxidant

K. Valachová, P. Rapta, M. Slováková, E. Priesolová, M. Nagy, D. Mislovičová, F. Dráfi, K. Bauerová, and L. Šoltés

CONTENTS

1.1 INTRODUCTION

Hyaluronan (HA, Figure 1) is a linear polysaccharide consisting of repeating disaccharide units of β-1,3-*N*-acetyl-D-glucosamine and β-1,4-D-glucuronic acid. The HA molar sizes in the human organism vary between 2×10^5 and 10×10^6 Da. An adult person weighing 70 kg has about 15 g of HA in the body, yet about one-third of this amount turns over daily.

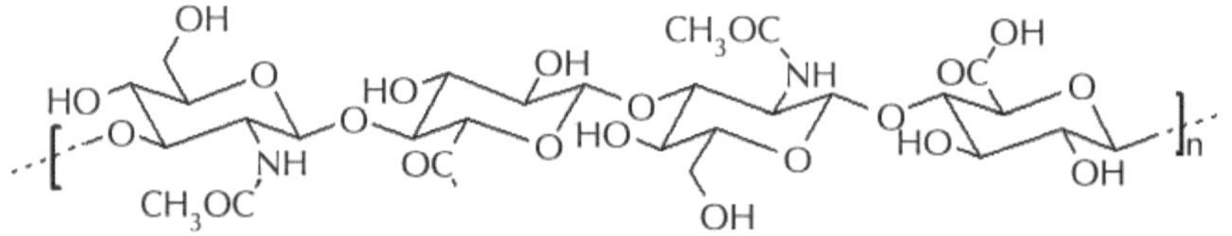

FIGURE 1 The HA acid form of the macromolecule.

In the human organism, HA is present in two forms: associated with certain glycosaminoglycans (GAGs) and proteins for example in the skin and cartilage, or free/unassociated in the synovial fluid (SF) and vitreous humor of the eye. The highest contents of (associated) HA in the human body occurs in the skin in which HA has a rapid turnover rate with a half-life of 1.5 days. Unassociated high molar mass HA in SF confers its unique viscoelastic properties required for maintaining proper functioning of the synovial joints. The half-life of HA within SF is approximately 12 hr [1].

The fast HA turnover in SF of the joints of healthy individuals may be attributed to the oxidative/degradative action of reactive oxygen species (ROS), generated among others by the catalytic effect of transition metal ions on the autoxidation of ascorbate [2]. Uninhibited and/or inhibited HA degradation by the action of various ROS has been studied on applying several *in vitro* models. In these studies, the change of the HA molar mass or a related parameter, such as the HA solution dynamic viscosity, was used as a marker of inflicted damage [3].

Arbutin (termed also hydroquinone-β-D-glucopyranoside, Figure 2), a substance used in the function of skin lightening and depigmentation is a popular drug especially in Japan and other Asian countries, also due to its lower toxicity compared to hydroquinone [4, 5]. Two anomeric forms, namely α- and β-arbutin, exist from which the latter occurs naturally in plants, for example in leaves of bearberry (*Arctostaphylos uva-ursi* Spreng., Ericaceae) and pear trees (*Pyrus communis* L., Rosaceae).

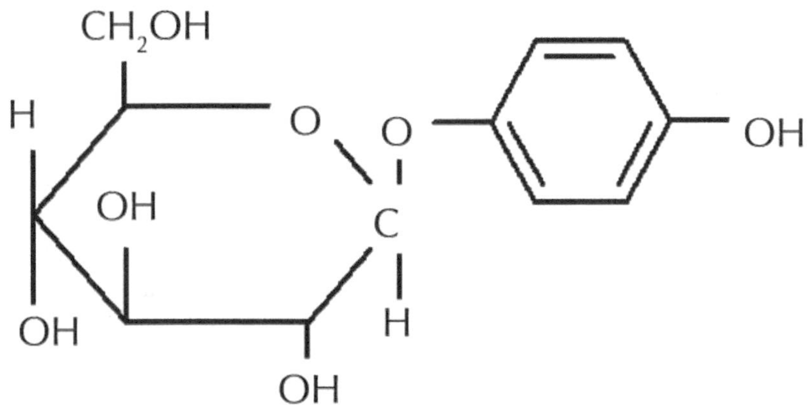

FIGURE 2 Arbutin structure.

The primary mechanism of arbutin action is in inhibiting tyrosinase activity in the skin, resulting in the significantly diminished formation of the brown colored melanin pigment. Further, arbutin acts as an "anti-aging" agent and a UVB/UVC filter which protects the skin against deleterious effects caused by photon generated free radicals. Beyond their cosmetical actions, products containing arbutin are therapeutically used for treating cystitis and kidney stones. Arbutin, a very hygroscopic substance, readily hydrolyzes for example by diluted acids yielding D-glucose and hydroquinone in a 1:1 mole ratio [6]. Arbutin based products are successfully applied also as diuretics: it is anticipated that the drug inhibits/retards the proliferation of pathogens within the urinary tract according to its high electrochemical potential, which within the pH range of 7.5 and 2.0 lies between +466 and +691 mV, respectively [7].

In the body, arbutin is decomposed to D-glucose and hydroquinone. It is a well known fact the latter substance bears strong antimicrobial and disinfectant properties. The greater the pH values of urine, the stronger the bacteria killing efficiency of free (not glucuronided) hydroquinone. However, in more acidic urine, hydroquinone undergoes extensive glucuronidation, which is believed to lead to the formation of the antibacterially ineffective hydroquinone glucuronide [7, 8].

So far no reports have been published devoted to action(s) of arbutin under conditions simulating ROS damaging effects on HA macromolecules. Since the HA content is so high in various tissues/body fluids–the skin/SF–the presented study is focused on investigating the ability of arbutin to act anti and/or prooxidatively by monitoring the kinetics of free radical degradation of high molar mass HA. As an effective $^{\bullet}OH$ radical inducer, the system comprising ascorbate plus Cu(II) – the so-called Weissberger's oxidative system was used. Along with monitoring the dynamic viscosity of HA solution by the method of rotational viscometry, further methods such as size exclusion chromatography (SEC), cyclic voltammetry (CV), EPR spectrometry, and standardized decolorization 2,2'-azino-bis(3-ethylbenzothiazoline-6-sulphonic acid) (ABTS) assay were applied at testing arbutin in the function of preventive and chain breaking antioxidant.

Arbutin was tested in the function of a potential anti or prooxidant in Cu(II) plus ascorbate induced degradation of high molar mass HA. The time and dose dependences of dynamic viscosity changes of the HA solutions were investigated by the method of rotational viscometry. First, the HA solution was exposed to degradation induced by Cu(II) ions (1 μM) with various concentrations of ascorbic acid (10, 50, or 100 μM). Further the action of arbutin (100 μM) addition into the reaction system, preceding that of ascorbic acid (100 μM), was inspected. The results obtained clearly indicated prooxidative properties of arbutin in relation to HA free radical degradation. Both the reduction of the dynamic viscosity of the HA solution and the decrease of the polymer mean molar mass as revealed by the method of SEC proved the tenet that on using the Cu(II) ions plus ascorbate, that is the Weissberger's oxidative system, the degradation of HA macromolecules is pronounced by added arbutin. Cyclovoltammetric studies of arbutin both in aqueous (0.15 M NaCl) and nonaqueous (0.15M TBAPF6 in DMSO) solutions confirmed a substantial protonation and hydrolysis reactions of arbutin in aqueous solutions. By applying the method of Electron Paramagnetic Resonance (EPR) spectroscopy, the reaction mixture comprising HA, Cu(II) ions, ascorbate, and arbutin indicated the presence exclusively of ascorbyl radicals. A remarkable radical scavenging activity of arbutin was observed on using the ABTS decolorization assay.

1.2 MATERIAL AND METHODS

1.2.1 Biopolymer

The high molar mass HA sample P9710-2A (M_w = 808.7 kDa; M_w/M_n = 1.63) used was obtained from Lifecore Biomedical Inc., Chaska, MN, U.S.A. The declared contents of transition metals in the HA sample given by the Certificate of Analysis are 4 ppm Cu and 13 ppm Fe.

1.2.2 Chemicals

Analytical purity grade NaCl and $CuCl_2{\cdot}2H_2O$ were purchased from Slavus Ltd., Bratislava, Slovakia. Dimethyl sulfoxide (DMSO), L-ascorbic acid, and $K_2S_2O_8$ were the products of Merck KGaA, Darmstadt, Germany. Trolox, arbutin, and 5,5-dimethyl-1-pyrroline-N-oxide (DMPO) were from Sigma-Aldrich Chemie GmbH, Steinheim, Germany. Tetrabutylammonium perchlorate (TBAP), $LiClO_4$, and ABTS (purum, >99%) were from Fluka, Chemie GmbH, Steinheim, and Germany. Redistilled deionized high quality grade water, with conductivity of <0.055 mS/cm, was produced by using the TKA water purification system from Water Purification Systems GmbH, Niederelbert, Germany.

1.2.3 Preparation of Stock and Working Solutions

The HA sample solutions (2.5 mg/ml) were prepared in the dark at room temperature in 0.15M aqueous NaCl in two steps: First, 4.0 ml of the solvent was added to 20 mg of dry HA powder. Then 3.80, 3.85, or 3.90 ml of the solvent was added after 6 hr. The stock solutions of L-ascorbic acid (1.6, 8, and 16 mM), arbutin (16 mM), and $CuCl_2$ (160 μM prepared from 16 mM $CuCl_2$) were also prepared in 0.15M aqueous NaCl.

1.2.4 Study of Uninhibited/Inhibited HA Degradation

Uninhibited

The HA degradation was induced by the oxidative system comprising $CuCl_2$ (1.0 μM) with altering concentrations of L-ascorbic acid (10, 50, and 100 μM). The procedure was as follows: the volume of 50 μl of 160 μM $CuCl_2$ solution was added to the HA solution (7.90 ml) and after 30 s stirring the reaction solution was left to stand for 7.5 min at room temperature. Then 50 μl of L-ascorbic acid (1.6, 8, or 16 mM) were added to the reaction vessel and the solution was gently stirred for 30 s. The reaction mixture was then immediately transferred into the viscometer Teflon cup reservoir.

Inhibited

The HA degradation: The procedures to test arbutin in the function of (i) preventive and (ii) chain breaking antioxidant were as follows:

(i) A volume of 50 μl of 160 μM $CuCl_2$ solution was added to the HA solution (7.85 ml) and after 30 s stirring the reaction solution was left to stand for 7.5 min at room temperature. Then 50 μl of arbutin (16 mM) were added to the solution and stirred again for 30 s. Finally, 50 μl of L-ascorbic acid (16 mM) were added to the reaction vessel and the solution was gently stirred for 30 s. The reaction mixture was then immediately transferred into the viscometer Teflon cup reservoir.

(ii) A similar procedure as described (i) was applied, however, 50 μl of L-ascorbic acid (16 mM) were added to the 7.5 min equilibrated reaction solution (7.90 or 7.85 ml)–comprising HA plus $CuCl_2$–and 30 s solution stirring followed. After 1 hr, finally 50 or 100 μl of arbutin (16 mM) were added to the solution and stirred again for 30 s. The solution mixture was then immediately transferred into the viscometer Teflon cup reservoir.

1.2.5 Rotational Viscometry

The resulting reaction mixture (8.0 ml) was transferred into the Teflon cup reservoir of a Brookfield LVDV-II+PRO digital rotational viscometer (Brookfield Engineering Labs., Inc., Middleboro, MA, U.S.A.). The recording of viscometer output parameters started 2 min after the experiment onset. The changes of the dynamic viscosity (η) values of the reaction mixture were measured at 25.0 ± 0.1°C in 3 min intervals for up to 5 hr. The viscometer Teflon spindle rotated at 180 rpm that is at a shear rate of 237.6 s^{-1}.

1.2.6 SEC Analyses

Evaluation of HA molar mass changes was performed with Shimadzu apparatus using a packed HEMA-BIO 1,000 column of dimensions 7.8 mm × 250 mm, the packing particle size was 10 μm. The mobile phase 100 mM phosphate buffer (pH 7.0) containing 0.15M NaCl was pumped by the LC-10AD device at a flow rate of 0.5 ml/min. For calibration of the SEC system used reference HAs (M_w = 90.2–1380 kDa) with broader molar mass distributions (M_w/M_n = 1.60–1.88) were applied. The samples of the volume of 20 μl with the polymer concentrations 1 mg/ml or lower were injected by a 7725i-type Rheodyne valve. The SEC analyses performance was monitored online by UV (SPD-10AV, set at 206 nm) and refractive index (RID-10) detectors.

1.2.7 Cyclic Voltammetry

All cyclovoltammetric experiments were performed with HEKA PG 284 (Germany) potentiostat under argon using a standard three-electrode arrangement of a platinum wire as working electrode, a platinum coil as counter electrode, and a Ag/AgCl as reference electrode. The scan rate used was 100 mV/s.

1.2.8 EPR Spectroscopy

The generation of free radicals during HA degradation was examined by spin trapping technique using an EPR X-band EMX spectrometer (Bruker, Rheinstetten, Germany) at ambient temperature [9, 10]. The sample preparation conditions were identical to those described under the paragraph "Study of uninhibited/inhibited HA degradation"–see the section "Uninhibited" as well as that of (i) within the section "Inhibited HA degradation". Yet the prepared total sample volume was only 800 μl and the addition of L-ascorbic acid solution (the last reactant) when appropriate was replaced by addition of the aqueous NaCl diluent (5 μL).

The volume of 250 μl of each sample solution was thoroughly mixed with 5.0 μl of DMPO spin trap prior to each experimental set carried out in a thin flat EPR quartz cell. The operational parameters of the equipment were adjusted as follows: Centre field 3354 G, sweep width 100 G, time constant 81.92 ms, conversion time 20.48 ms, receiver gain 5e + 5, microwave power 10 mW, and modulation amplitude 2 G.

1.2.9 ABTS Assay–Determination of the TEAC Value

The $ABTS^{\bullet+}$ radical cations were performed by the reaction of an aqueous solution of $K_2S_2O_8$ (3.3 mg) in water (5 ml) with ABTS (17.2 mg). The resulting bluish green radical cation solution was stored overnight in the dark below 0°C. Before experiment, the solution (1 ml) was diluted into a final volume (60 ml) with water [11].

Arbutin solution was prepared as specified under the paragraph "Preparation of stock and working solutions". The investigated samples comprised 2 ml of the diluted $ABTS^{\bullet+}$ solution with addition of 50 μl of working arbutin solution (1 mM).

A modified ABTS assay was run applying a UV/VIS S2000 spectrophotometer (Sentronic, Germany). The UV/VIS spectra were taken up to 120 min. The relative differences (ΔA) in the light absorbance at 731 nm in the 10th and 120th min compared to the reference experiment using Trolox were used to calculate the values of Trolox Equivalent of Antioxidant Capacity (TEAC).

1.2.10 ABTS Assay–Determination of the IC_{50} Value

The $ABTS^{\bullet+}$ radical cation solution (250 μl) prepared from 7 mM ABTS and 2.45 mM $K_2S_2O_8$, 1:1 v/v ratio of water solutions was added to 2.5 μl of the arbutin solution and the absorbance of the sample mixture was measured at 734 nm after 6 min. The radical scavenging capacity of arbutin was investigated at the drug concentration range 2–1000 μM.

The light absorbance was measured quadruplicately in 96-well Greiner UV-Star microplates (Greiner-Bio-One GmbH, Germany) with Tecan Infinite M 200 reader (Tecan AG, Austria). The IC_{50} value was calculated with CompuSyn 1.0.1 software (ComboSyn, Inc., Paramus, USA).

1.3 DISCUSSION AND RESULTS

1.3.1 Establishing a Proper HA Oxidative Degradation System

The HA macromolecules were exposed to free radical degradation induced by Cu(II) ions (1 μM) and ascorbate at various concentrations (10, 50, or 100 μM) where the primary variable—the concentration of cupric ions – was set to such a value which may occur at the early stages of acute joint inflammation. The second variable, the concentration of ascorbate, covers that range which is characteristic in SF of a human being. As shown in Figure 3, the curve coded 1, the reaction system Cu(II):ascorbate equaling 1:10 was during the whole time period of 5 hr ineffective to induce a relevant HA degradation. The observed minor growing of the solution η value represents the well known effect called rheopexy–which is characterized by orientations and attractive self-interactions of dissolved HA macromolecules in the space between the viscometer reservoir wall and the rotating spindle [2].

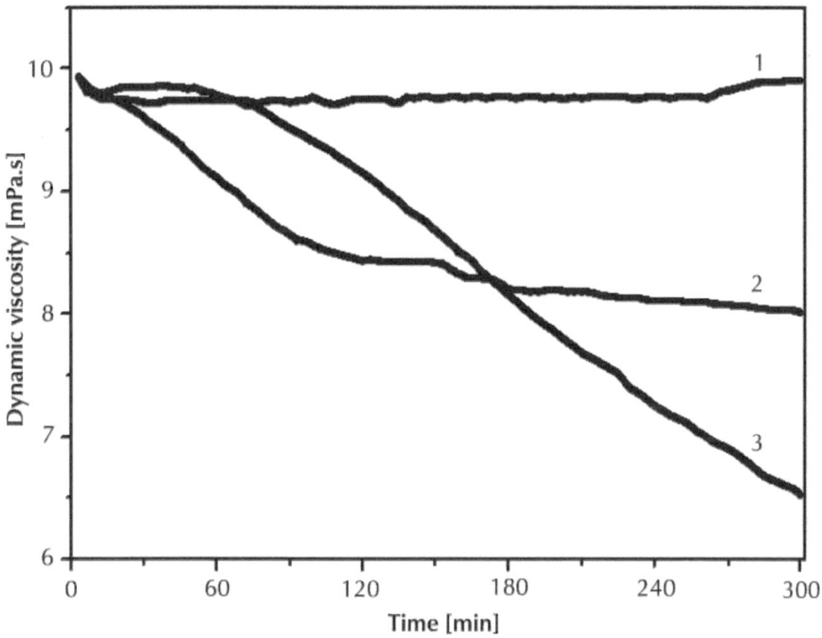

FIGURE 3 Time dependences of dynamic viscosity values of the HA solution in the presence of 1.0 μM $CuCl_2$ plus ascorbate at concentrations 10 (1), 50 (2), or 100 (3) μM.

By increasing the ascorbate concentration to 50 μM (compare Figure 3, curve coded 2), a gradual decrease of the dynamic viscosity of HA solution was evidenced during the first 90 min. Then a slower HA degrading process was observed, which continued up to the end of monitoring the solution η values 5 hr. The attained decrease of the HA dynamic viscosity from its initial value of 9.93 to 8.01 MPa·s was, however, classified as not sufficiently effective for further investigations.

Increasing the ascorbate concentration to 100 μM resulted in an approximately 60 min time interval during which "no degradation" of HA macromolecules was registered. Yet as evident from the curve coded 3 in Figure 3, this stationary/lag phase was continued by a stepwise decrease of the solution dynamic viscosity, achieving at 5 hr the value of 6.53 MPa·s. As proven by using the method of EPR spectroscopy [12] under the experimental conditions represented in Figure 3, curve coded 3, the signal detected during the earlier time intervals (≤ 60 min) belonged exclusively to ascorbyl radical. However as confirmed by applying the DMPO spin trapping method, at later time intervals, the generated $^{\bullet}OH$ radicals "exceeded" the scavenging effectivity of the actual ascorbate ($AscH^-$) level and thus they represented the radical species which reacted with the HA chains leading to polysaccharide fragmentation. As seen in Figure 3, curve coded 3, during the time interval between 60 and 300 min the continual stepwise decrease of the solution η values covers a range equaling to 3.25 MPa·s.

Earliest Time Interval of the Reaction

Let us first analyze the potential chemical processes which resulted in the observations represented by the curve coded 3 in Figure 3 namely the time interval up to approximately 60 min: Under aerobic conditions, the bicomponent system comprising ascorbate and a trace amount of cupric ions is a well known Weissberger's oxidative system generating the strong oxidant hydrogen peroxide. The following flow chart can be primarily used to characterize the individual reaction steps which however take place simultaneously in a concerted action.

$$AscH^- + Cu(II) \rightarrow Asc^{\bullet-} + H^+ + Cu(I) \text{ (reduction of cupric to cuprous ions)} \quad (1)$$

Note 1: It is a well known fact that $AscH^-$ donates a hydrogen atom ($H^{\bullet}$ or $H^+ + e^-$) yielding the resonance stabilized tricarbonyl ascorbate free radical. Since the latter has a pK_a value of 0.86, it is not protonated and thus in aqueous solution it will be present as $Asc^{\bullet-}$ [13].

As the cuprous ions are unstable, namely they can quickly undergo a disproportional reaction $Cu(I) + Cu(I) \rightarrow Cu(0) + Cu(II)$, they are charge stabilized by an excess of the present ascorbate as follows $AscH^- + Cu(I) \rightarrow [AscH^- \cdots Cu(I)]$. This reaction intermediate participates in the next bielectron reduction of the dioxygen molecule $(O = O)$

$$[AscH^- \cdots Cu(I)] + O{=}O + H^+ \rightarrow Asc^{\bullet-} + Cu(II) + H_2O_2 \text{ .(hydrogen peroxide formation)} \quad (2)$$

and one may suppose that the nascent H_2O_2 molecule is decomposed immediately by the "uncomplexed"/complexed cuprous ion(s) according to a Fenton-like reaction:

$$H_2O_2 + [Cu(I)] \rightarrow {}^{\bullet}OH + HO^- + Cu(II) \text{ (generation of hydroxyl radicals)} \quad (3)$$

In summary, we could express the following final/net reaction:

$$AscH^- + [AscH^- \cdots Cu(I)] + O{=}O \rightarrow 2Asc^{\bullet-} + Cu(II) + HO^- + {}^{\bullet}OH \text{ .(net reaction)}$$

Yet the conditions of the flow chart should be taken into account: Since within the reaction system there is a real ascorbate excess, the nascent hydroxyl radicals are continually quenched and due to these facts the EPR spectrometer monitors exclusively the ascorbyl radicals.

$$AscH^- + {}^{\bullet}OH \rightarrow Asc^{\bullet-} + H_2O \text{ (quenching the hydroxyl radicals)} \quad (4)$$

According to our findings obtained by using the method of EPR spectroscopy [12] under the experimental conditions represented also in Figure 3, curve coded 3, during the earlier time intervals (≤ 60 min) the monitored ascorbyls – $Asc^{\bullet-}$–indicate that any potentially generated hydroxyl radical is quenched immediately at its phase of nascence.

The informed reader knows that (two) ascorbyls undergo a redox disproportional reaction yielding back a molecule of ascorbate and a not charged dehydroascorbate (DHA). The DHA molecule however hydrolyzes yielding an intermediate, namely 2,3-diketo-L-gluconic acid, and/or several final products such as L-xylose, and L-xylonic-, L-lyxonic-, L-threonic-, as well as oxalic acids [13]. Study of the participation of the mentioned intermediate as well as of the final products within the reaction flow charts (compare reactions 1–4) was however beyond the scope of the present investigation.

Later Time Interval of the Reaction

Let us continue the analysis of the potential chemical processes which could result in the observations represented by the curve coded 3 in Figure 3 namely the result within the time interval from approximately 60 to 300 min: As anticipated according to the sequential depletion of the ascorbate the nonquenched hydroxyl radicals can react with the HA macromolecule by abstracting a hydrogen radical ($H^{\bullet}$), resulting in the formation of a *C*-macroradical represented in Figure 4 and further denoted as $A^{\bullet}$.

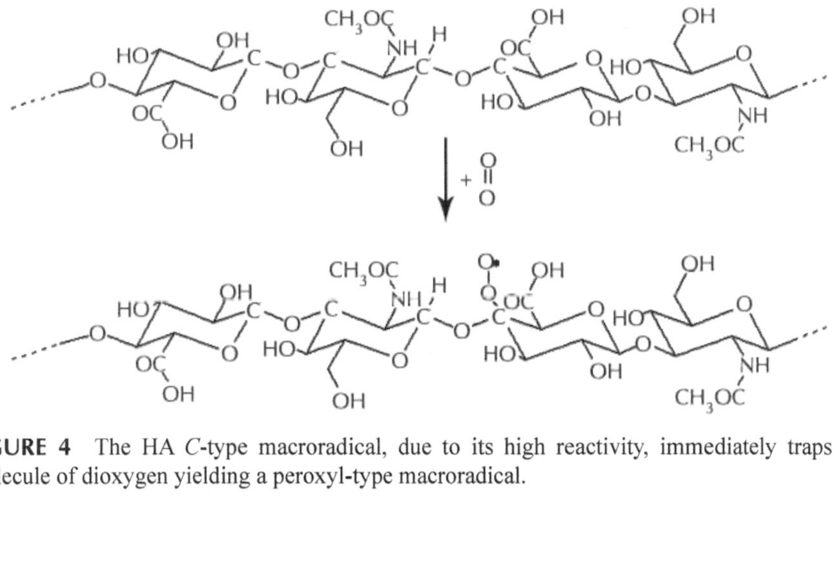

FIGURE 4 The HA *C*-type macroradical, due to its high reactivity, immediately traps a molecule of dioxygen yielding a peroxyl-type macroradical.

Under aerobic conditions, the alkyl-type macroradical – A• – reacts rapidly with the molecule of dioxygen (O_2) yielding a peroxyl type macroradical, hereafter denoted as AOO•. The intermediate peroxyl type macroradical formed may react with an adjacent HA macromolecule (HA), and thus the radical chain reaction propagates quickly.

$$AOO^{\bullet} + HA \rightarrow AOOH + A^{\bullet} \text{ (propagation of the radical chain reaction)} \quad (5)$$

After its "collision" with an HA macromolecule (compare reaction 5), the generated peroxyl type macroradical yields a high molar mass hydroperoxide which subsequently, mostly induced by the present Cu(I) ions {$AOOH + Cu(I) \rightarrow AO^{\bullet} + HO^{-} + Cu(II)$}, yields an alkoxyl type macroradical (AO•). This is a presumed intermediate of the main chain splitting, resulting in biopolymer fragments whose solution is characterized by a reduced dynamic viscosity (compare Scheme 1 [14]).

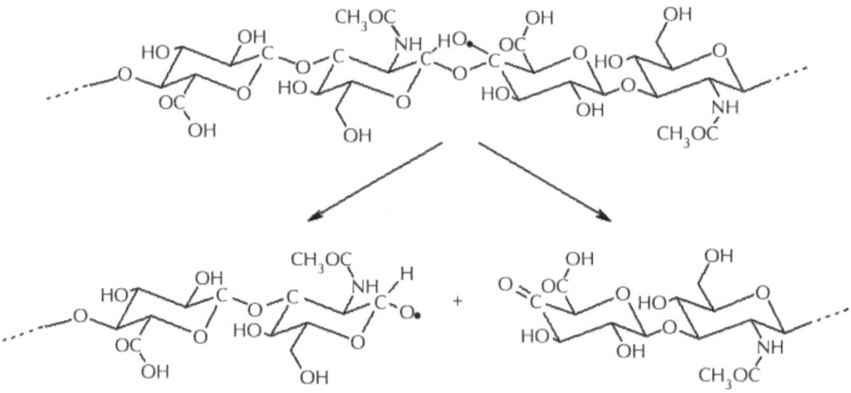

SCHEME 1 Fragmentation of the alkoxy type macroradical (AO•).

The attack of hydroxyl radicals on D-glucuronate or *N*-acetyl-D-glucosamine moieties of HA can also lead to the "opening" of rings without breaking the polymer chain [15, 16].

1.3.2 Investigating Pro and Antioxidative Action of Arbutin

Based on its chemical structure (compare Figure 2), the natural compound arbutin can be grouped among O-alkyl derivatives of hydroquinone which are generally classified as efficient antioxidants [17]. To assay arbutin action in its function as a preventive antioxidant, the drug was loaded into cupric ions containing HA solution, directly before ascorbic acid admixing as described in procedure (i) under the paragraph "Study of uninhibited/inhibited HA degradation". As indicated in Figure 5, contrary to the expected inhibition/retardation of the degradation of HA macromolecules, on introducing the drug – resulting in its concentration of 50 or 100 μM (compare curves coded 2' or 3' in Figure 5) – a remarkable reduction of the η value of the solution monitored was recorded at the earliest time interval. While the shape of curve 3 (compare Figure 5), relating to the reaction mixture without arbutin, indicates a stationary/lag phase of the HA degradation process during the first 60 min, addition of the arbutin drug tested (100 μM) shows both elimination of the reaction lag phase and significant acceleration of the decay of HA macromolecules.

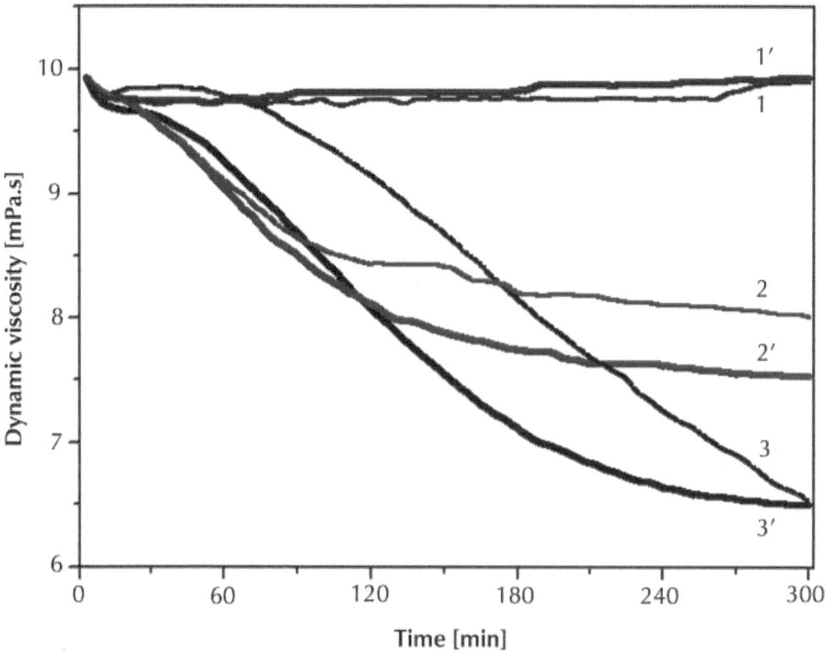

FIGURE 5 Time dependences of dynamic viscosity values of the HA solution in the presence of 1 μM $CuCl_2$, 100 μM arbutin, plus ascorbate at the concentrations 10 (1'), 50 (2'), or 100 (3') μM, as well the dependences observed under similar experimental condition without any arbutin – curves 1, 2, and 3 – relate to ascorbate concentrations of 10, 50, and 100 μM, respectively.

The degradation of the native HA sample after 5 hr treatment by the system comprising Cu(II):arbutin:ascorbate in the molar concentration ratio of 1:100:100 was unequivocally proved by SEC analysis (compare Figure 6, UV and RI traces). As evident from Figure 6, the polymer sample at 5 hr had a lower molar mass ($M_{peak} \approx$ 695 kDa) as compared to that of the original HA.

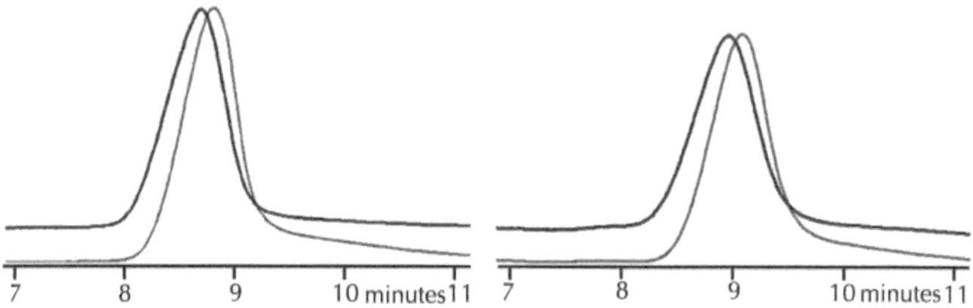

FIGURE 6 The SEC analyses of the native HA sample and of that degrading over 5 hr. (The native HA peaked earlier.) Left panel – UV traces at 206 nm; right panel – RI records.

To explain the observations, the reader's attention is to be switched to the reversible redox processes known for the pair of hydroquinone and quinone (denoted hereafter shortly QH_2 and Q), which can be described as follows [18]:

$$QH_2 \leftrightarrow Q + 2H^+ + 2e^- \text{ (hydroquinone two-electron oxidation)}$$

As claimed, arbutin is readily hydrolyzed by diluted acids yielding QH_2 and D-glucose in a 1:1 mole ratio [6]. The latter tenet was supported also by cyclic voltammetric experiments performed both in DMSO/TBAPF6 and aqueous 0.15M NaCl solution, as shown in Figure 7. Clear differences are seen on comparing DMSO and water solutions. In water solution, the products of hydrolysis as well as protonation of arbutin are observable already during the first CV scan, in contrast to DMSO where only a small increase of arbutin degradation products is seen. In water solution, the strong increase of the CV peak, characteristic of Q/QH_2 redox couple during repetitive redox cycling, indicates that compared to non-oxidized arbutin, the oxidized form of arbutin strongly hydrolyzes (decomposes). Thus due to the addition of ascorbic acid to arbutin, rapid drug hydrolysis occurred.

Reaction Chemistry Investigated by Spin-trapping EPR Spectroscopic Method

Figure 8 represents typical records registered on applying DMPO in the function of radical trapping within the reaction system comprising (a) HA and Cu(II), (b) HA, Cu(II), and arbutin, as well as (c) HA, Cu(II), arbutin, and ascorbate, where the reactants were applied at such concentration ratios which were identical to those described under the paragraph "Study of uninhibited/inhibited HA degradation".

(a, b): By admixing DMPO into the reaction vessel containing already the two components – HA and Cu(II) – a signal characteristic for the $^{\bullet}$DMPO-OH adduct was recorded, however the following arbutin application diminished somewhat the signal abundance (compare Figure 8, traces coded **a** and **b**). The simplest explanation of these observations could be that DMPO added in a great excess reduced Cu(II) ions to cuprous ones, which subsequently reacted with dioxygen, yielding superoxide anion radicals ($O_2^{\bullet-}$). These ROS spontaneously dismutate and the dismutation product H_2O_2 is decomposed by the reduced copper ions in a Fenton-like reaction.

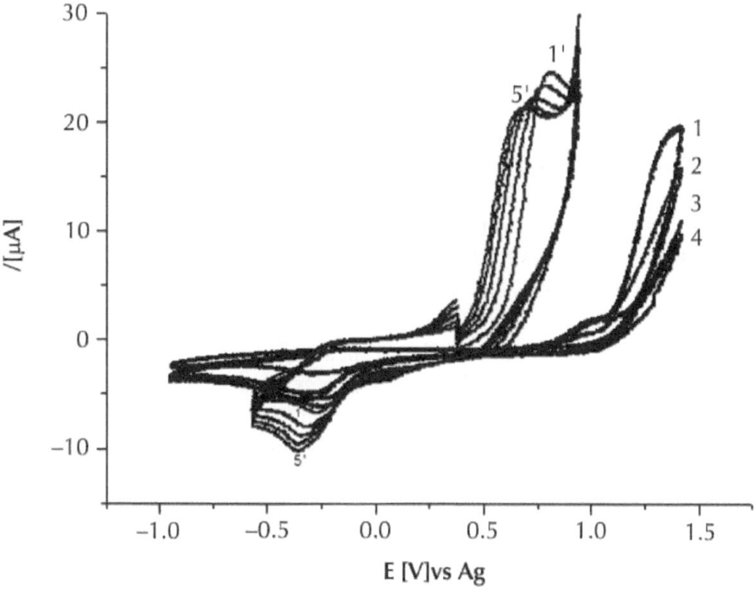

FIGURE 7 The CV (repetitive cycling, CV scans are indicated with numbers) of arbutin in DMSO/TBAPF6 (black lines 1–4) or in H_2O/NaCl (gray lines 1'–5').

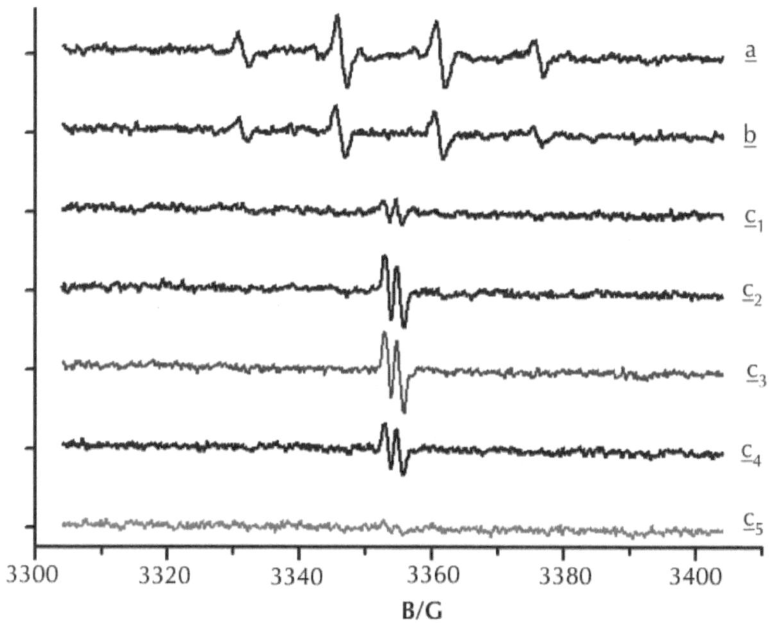

FIGURE 8 The ˙DMPO-OH signals registered 1 min after mixing the DMPO spin trap with the sample solution containing HA and Cu(II) – trace **a**; and with the solution comprising HA, Cu(II), and arbutin – trace **b**.

Note 2: It should be pointed out, however, that systems comprising HA and Cu(II) or HA, Cu(II), and arbutin did not induce any significant degradation of HA macromolecules (not shown).

The signals for ascorbyl detected at time intervals of 10, 30, 60, 90, and 120 min are represented by the respective traces from $\underline{\mathbf{c}}_1$ to $\underline{\mathbf{c}}_5$.

(c): It should be pointed out here that the mixing of all the reaction components, namely DMPO with the sample solution containing HA, Cu(II), arbutin, and ascorbate, took but approximately 1 min during which time the sample was not scanned by the EPR spectrometer. However, after a short time, a signal related to $Asc^{\bullet-}$ emerged, whose intensity varied: During the first time interval up to approximately 60 min the intensity of the recorded signal rose, then it declined and after 120 min the spectrometer monitor was "silent" (compare Figure 8, traces $\underline{\mathbf{c}}_1$–$\underline{\mathbf{c}}_5$). The EPR spectroscopic observations at a later time interval, namely the less and less abundant signal related to $Asc^{\bullet-}$ between 60 and 120 min, as well as later on the signal disappearance, coincide well with the results represented in Figure 5, curve coded 3' at similar time intervals.

By admixing DMPO into the reaction vessel containing already the two components – HA and Cu(II) – a signal characteristic for the •DMPO-OH adduct was recorded, however the following arbutin application reduced the EPR signal of •DMPO-OH adducts by about 20–30% (compare Figure 9).

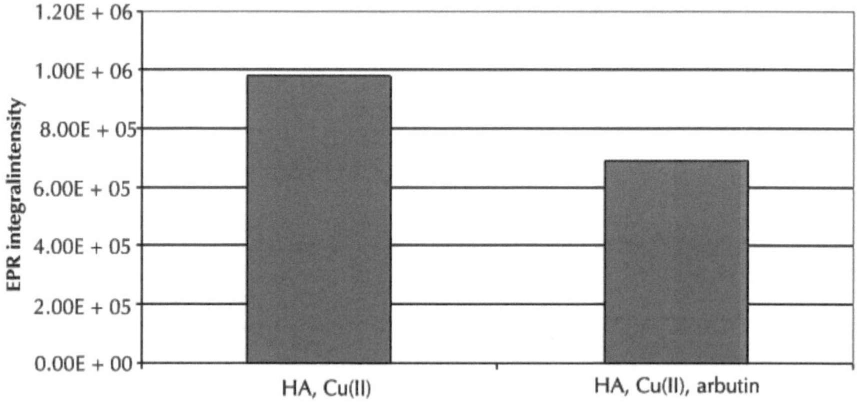

FIGURE 9 Integral EPR intensity of •DMPO-OH adducts registered 1 min after mixing the DMPO spin trap with sample solution containing HA and Cu(II) and with the solution comprising HA, Cu(II), and arbutin.

The simplest explanation of these observations could be that hydroperoxides already present in the initial HA solution [9, 19, 20] rapidly decompose due to the presence of Cu(II)/Cu(I) ions (compare also Scheme 1), as already well established in the field of peroxidation of lipids (denoted here as L)

$$LOOH + Cu(II) \rightarrow LOO^{\bullet} + Cu(I) + H^{+}$$
$$LOOH + Cu(I) \rightarrow LO^{\bullet} + Cu(II) + HO^{-}$$

Note 3: It is well known that the presence of peroxyl and alkoxyl radicals as well as their DMPO spin adducts in water solutions leads mostly to the formation of $^{\bullet}OH$ radicals generating the signal of $^{\bullet}$DMPO-OH adduct. Arbutin is probably able to eliminate reactive alkoxy and peroxyl radicals before their spin trapping.

In the case of the system comprising a trace amount of oxidized HA, or more precisely the HA sample containing already a preformed macrohydroperoxides (AOOH), the two electrons gained according to reaction (6) could reduce Cu(II) ions to Cu(I). The generated cuprous ions are then involved in the following reaction

$$AOOH + Cu(I) \rightarrow AO^{\bullet} + Cu(II) + HO^{-}$$

Thus a short-living (ca. 1 μs) alkoxy type macroradical was only present in our experimental system. This macroradical was extremely unstable and degraded quickly to fragments with lower molar mass (compare Scheme 1).

Note 4: It should be pointed out, however, that systems comprising HA and Cu(II) or HA, Cu(II), and arbutin did not induce any significant degradation of HA macromolecules (not shown).

ABTS Assay of Arbutin

The determined TEAC values that is those for the 10th min 1.22 mmol Trolox/L and for the 120th min 1.59 mmol Trolox/L (compare Figure 10) prove a really high reductant power of arbutin in aqueous solutions.

The arbutin scavenging activity represented by its established IC_{50} value 5.43 mM should be classified as relatively high, comparable to that value of quercetin (IC_{50} = 2.86 mM) – the substance used as a standard effective natural antioxidant [21, 22].

The reaction of decolorization of the $ABTS^{\bullet+}$ radical cation solution is based on a simple reaction during which a proper reductant provides an electron to the acceptor – the $ABTS^{\bullet+}$ radical cation. Although the ABTS assay is still one of the primary choices, the reader should take into account that the results described by either the TEAC or IC_{50} values represent just values determined at the experimental conditions used by applying for example another electron acceptor (such as that of $DPPH^{\bullet}$ – the 1,1-diphenyl-2-picrylhydrazyl radical) different IC_{50} values are established.

Note 5: On applying the DPPH assay, the determined IC_{50} value for arbutin (≈ 768 mM) represents a 176 times lower antioxidant capacity than the one determined for quercetin (4.36 mM). Although the arbutin IC_{50} value correlates well with those published by other investigators applying similar DPPH-assay conditions [23], it should be pointed out that there is an experimental "conflict": The DPPH assay, which uses alcohols (methanol) as the 1,1-diphenyl-2-picrylhydrazyl-radical dissolvent, is preferably used to investigate the antioxidant capacity of lipophilic (alcohol soluble) substances. Yet arbutin is classifiable as a hydrophilic substance, perfectly soluble in water. As calculated, the arbutin *versus* quercetin solubility in water, 39.1 *versus* 0.26 g/l, closely correlates to the logP - lipophilicity values of these substances, namely -1.36 *versus*. + 1.07 [24], respectively. Thus, the IC_{50} value for arbutin (≈ 768 mM) determined by using the DPPH assay does not primarily indicate the very low reduction power of the drug, but most plausibly, this value reflects also a much poorer arbutin solubility compared to that of quercetin when methanol was the solvent used.

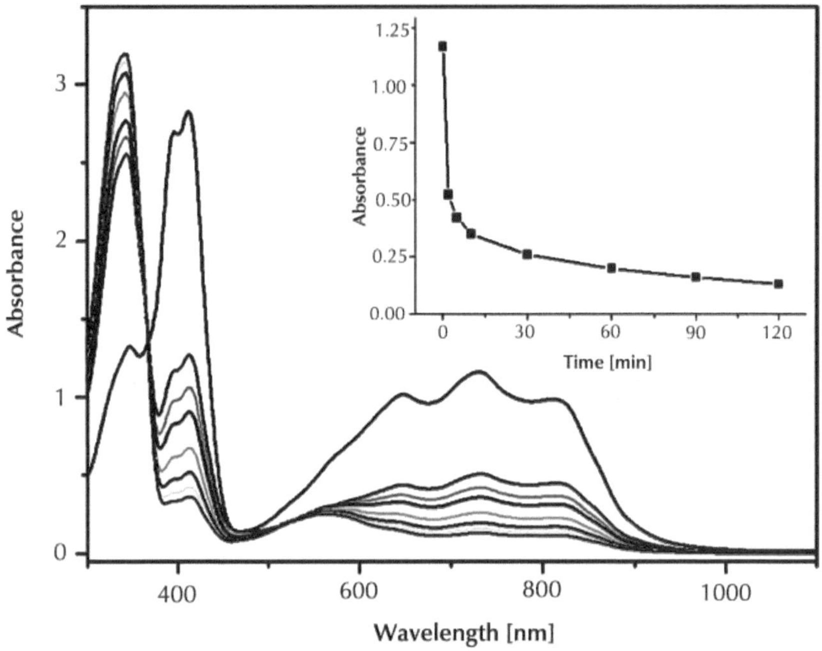

FIGURE 10 The UV/Vis records documenting arbutin scavenging of the $ABTS^{•+}$ radical cation. Inset: Time dependence of the absorbance at 731 nm.

1.3.3 Investigating Arbutin in the Function of a Chain Breaking Antioxidant

There is a discrepancy of arbutin classification as an efficient antioxidant/reductant based on the ABTS assays and on the contrary a pro-oxidant investigated by the drug action when introduced into the running free radical HA degradation during the phase of the reaction propagation. Under the experimental conditions specified under the paragraph "Study of uninhibited/inhibited HA degradation" (ii) the results represented in Figure 11 should be assessed as biphasic. As evident (compare Figure

11, curves coded 100 and 200), addition of the arbutin solution after 1 hr of the reaction onset, immediately increased the degradation rate of HA macromolecules. This process however continued by a stepwise slowing down of the solution dynamic viscosity with the final values 8.43 and 7.33 MPa·s, proving the inhibitory action of arbutin compared to uninhibited free-radical HA degradation represented by the curve coded 0 in Figure 11.

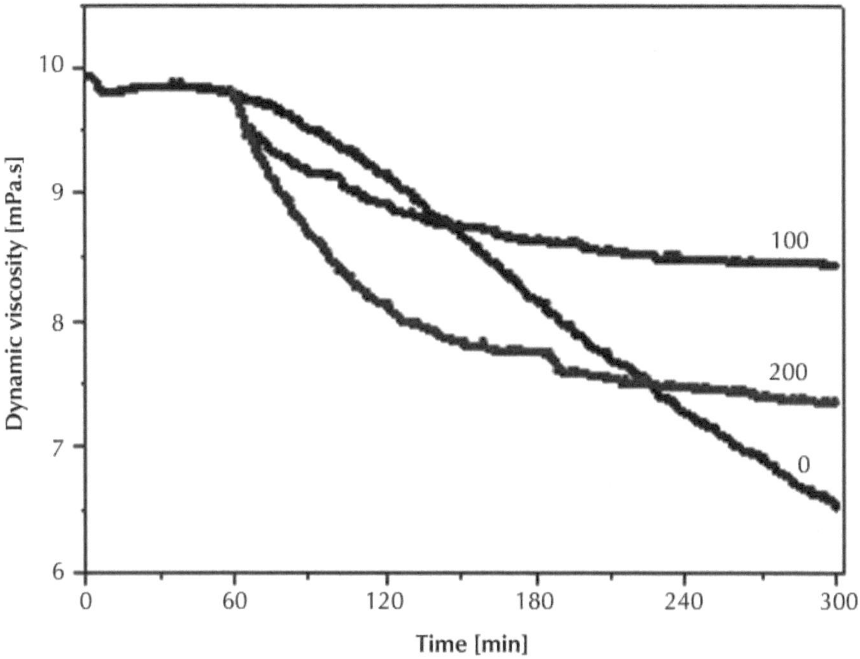

FIGURE 11 Time dependences of dynamic viscosity values of HA solution in the presence of 1 μM $CuCl_2$ plus 100 μM ascorbate (0) and after addition of arbutin in the concentration of 100 (100) and 200 (200) μM.

1.4 CONCLUSION

Based on the results (compare Figure 11), one may conclude that arbutin functions as a chain-breaking antioxidant. Yet it should be pointed out again that such a statement is valid only under the experimental conditions used. On applying the drug for skin lightening, one should take care since the pro-oxidative action of arbutin may result in damaging the tissues involved. The skin contents of HA, ascorbate, and the ubiquitous redox-active (transition) metals [18] including Cu ions predispose the processes documented by the observations.

KEYWORDS

- **ABTS assay**
- **Cyclic voltammetry**
- **EPR spectroscopy**
- **Hydroquinone derivatives**
- **Quinone**
- **Rotational viscometry**
- **Size exclusion chromatography**
- **Weissberger's oxidative system**

ACKNOWLEDGMENT

The work was supported by the VEGA grant project Nos.: 1/0145/10, 2/0011/11, and 2/0045/11 of the Slovak Academy of Sciences and by the grant projects APVV Nos.: 0488-07 and 0351-10.

The work was supported also by the Research and Development Operational Program of ERDF (project implementation: Centre of excellence for white-green biotechnology, ITMS 26220120054).

REFERENCES

1. Kogan, G., Šoltés, L., Stern, R., Schiller, J., and Mendichi, R. Hyaluronic acid: Its function and degradation in in vivo systems. *Studies in Natural Products Chemistry* (*Vol. 34 Bioactive Natural Products, Part D*). Atta-ur-Rahman (Ed.) Elsevier, Amsterdam, pp. 789–882 (2008).
2. Šoltés, L., Stankovská, M., Kogan, G., Gemeiner, P., and Stern, R. *Chem. Biodivers.*, **2**, 1242–1245 (2005).
3. Šoltés, L., Mendichi, R., Kogan, G., Schiller, J., Stankovská, M., and Arnhold, J. *Biomacromolecules*, **7**, 659–668 (2006).
4. http://ntp.niehs.nih.gov/NTP/Noms/Support_Docs/Hydroquinone_may2009.pdf
5. Hincha, D. K., Oliver, A. E., and Crowe, J. H. *Biophys. J.*, **77**, 2024–2034 (1999).
6. Budavari, S. *The Merck Index 12th ed.*, Merck and Company, Inc., CD-ROM version 12:1 Chapman and Hall Electronic Publishing Division, Whitehouse Station, New Jersey (1996).
7. Nycz, J. E., Malecki, G., Morag, M., Nowak, G., Ponikiewski, L., Kusz, J., and Switlicka, A. *J. Molec. Structure*, **980**, 13–17 (2010).
8. http://www.greatvistachemicals.com/proteins-sugars-nucleotides/arbutin.html
9. Hrabárová, E., Valachová, K., Rychlý, J., Rapta, P., Sasinková, V., Malíková, M., and Šoltés, L. *Polym. Degrad. Stabil.*, **94**, 1867–1875 (2009).
10. Valachová, K., Rapta, P., Kogan, G., Hrabárová, E., Gemeiner, P., and Šoltés, L. *Chem. Biodivers.*, **6**, 389–395 (2009).
11. Hrabárová, E., Valachová, K., Rapta, P., and Šoltés, L. *Chem. Biodivers.*, **7**, 2191–2200 (2010).
12. Šoltés, L., Stankovská, M., Brezová, V., Schiller, J., Arnhold, J., Kogan, G., and Gemeiner, P. *Carbohydr. Res.*, **341**, 2826–2834 (2006).
13. Buettner, G. R. and Schafer, F. Q. Ascorbate (Vitamin C), its Antioxidant Chemistry. The Virtual Free Radical School for Oxygen Society Powerpoint presentation. http://www.healthcare.uiowa.edu/corefacilities/esr/publications/buettnerpubs/pdf/Buettner-Ascorbate-Chemistry-1.pdf (accessed on 2002)
14. Rychlý, J., Šoltés, L., Stankovská, M., Janigová, I., Csomorová, K., Sasinková, V., Kogan, G., and Gemeiner, P. *Polym. Degrad. Stabil.*, **91**, 3174–3184 (2006).

15. Hawkins, C. L. and Davies, M. J. *Biochem. Soc. Trans.*, **23**, 248(1995).
16. Hawkins, C. L. and Davies, M. J. *Free Radic. Biol. Med.*, **21**, 275–290 (1996).
17. http://monographs.iarc.fr/ENG/Monographs/vol71/mono71-30.pdf
18. Song, Y. and Buettner, G. R. *Free Radic. Biol. Med.*, **49**, 919–962 (2010).
19. Stankovská, M., Arnhold, J., Rychlý, J., Spalteholz, H., Gemeiner, P., and Šoltés, L. *Polym. Degrad. Stabil.*, **92**, 644–652 (2007).
20. Šoltés, L., Kogan, G., Stankovská, M., Mendichi, R., Rychlý, J., Schiller, J., and Gemeiner, P. *Biomacromolecules*, **8**, 2697–2705 (2007).
21. Jančinová, V., Petríková, M., Perečko, T., Drábiková, K., Nosáľ, R., Bauerová, K., Poništ, S., and Košťálová, D. *Chemické listy – Chem. Letters*, **101**, 189–191 (2007).
22. Pečivová, J., Mačičková, T., Jančinová, V., Drábiková, K., Perečko, T., Lojek, A., Číž, M., Carbolová, J., Košťálová, D., Cupaníková, D., and Nosáľ, R. *Interdiscipl. Toxicol.*, **2**, 137 (2009).
23. Takebayashi, J., Ishii, R., Chen, J., Matsumoto, T., and Ishimi, Y. Tai. *Free Radical Res.*, **44**, 473–478 (2010).
24. http://www.vcclab.org/lab/alogps/start.html

2 Diffusion of Electrolytes and Non-electrolytes in Aqueous Solutions: A Useful Strategy for Structural Interpretation of Chemical Systems

Ana C. F. Ribeiro, Joaquim J. S. Natividade, Artur J. M. Valente, and Victor M. M. Lobo

CONTENTS

2.1 INTRODUCTION

In the last years, the diffusion Coimbra group, headed by Prof. Lobo, has been particularly dedicated to the study of mutual diffusion behavior of binary, ternary, and quaternary solutions [1-19], involving electrolytes and non-electrolytes, helping to go deeply in the understanding of their structure, and aiming at practical applications in fields as diverse as corrosion studies occurring in biological systems or therapeutic uses. In fact, the scarcity of diffusion coefficients data in the scientific literature, due to the difficulty of their accurate experimental measurement and impracticability of their determination by theoretical procedures, allied to their industrial and research need, well justify our efforts in accurate measurements of such transport property.

This transport property has been measured in different conditions (several electrolytes, concentrations, temperatures, techniques used), having in mind a contribution to a better understanding of the structure of those solutions, behavior of electrolytes or non-electrolytes in solution and last but not the least, supplying the scientific and technological communities with data on this important parameter in solution transport processes. Whereas an open-ended capillary cell developed by Lobo has been used to obtain mutual diffusion coefficients of a wide variety of electrolytes [1, 2], the Taylor technique has been used mainly for ternary and quaternary systems with non-electrolytes (e.g., [11-19]). From comparison between experimental results and those obtained from different models, for example, Nernst, Nernst-Hartley, Stokes, Onsager and Fuoss, and Pikal theoretical equations, from semi-empirical equations, and Gordon's and Agar's as well, it has been possible to obtain some structural information, such as diffusion coefficient at infinitesimal concentration, ion association, complex formation, hydrolysis, hydration, and estimations of the mean distance of closest approach involving ions as diffusing entities.

2.1.1 Concepts of Diffusion in Solutions

Many techniques are used to study diffusion in aqueous solutions. It is very common to find misunderstandings concerning the meaning of a parameter, frequently just denoted by *D* and merely called diffusion coefficient, in the scientific literature, communications, meetings, and simple discussions among researchers. In fact, it is necessary to distinguish self-diffusion (interdiffusion, tracer diffusion, single ion diffusion, and ionic diffusion) and mutual diffusion (interdiffusion, concentration diffusion, and salt diffusion) [20, 21]. Nuclear magnetic resonance (NMR) and capillary tube, the most popular methods, can only be used to measure interdiffusion coefficients [20, 21]. In our case, the mutual diffusion is analyzed [20, 21].

Mutual diffusion coefficient, *D,* in a binary system, may be defined in terms of the concentration gradient by a phenomenological relationship, known as Fick´s first law.

$$J = -D\frac{\partial c}{\partial x} \tag{1}$$

where J represents the flow of matter across a suitable chosen reference plane per area unit and per time unit, in a one-dimensional system, and c is the concentration of solute in moles per volume unit at the point considered, Equation (1) may be used to measure D. The diffusion coefficient may also be measured considering Fick's second law.

$$\frac{\partial c}{\partial t} = \frac{\partial}{\partial x}\left(D\frac{\partial c}{\partial x}\right) \tag{2}$$

In general, the available methods are assembled into two groups: steady and unsteady-state methods, according to Equations (1) and (2). In most of the processes, diffusion is a three-dimensional phenomenon. However, many of the experimental methods used to analyze diffusion restrict it to a one-dimensional process, making it much easier to study its mathematical treatments in one dimension (which then may be generalized to a three-dimensional space).

The resolution of Equation (2) for a unidimensional process is much easier if we consider D as a constant. This approximation is applicable only when there are small differences of concentration, which is the case of open-ended conductometric technique and of the Taylor technique [20, 21]. In these conditions, it is legitimate to consider that measurements of differential diffusion coefficients obtained by the techniques are parameters with a well defined thermodynamic meaning [20, 21].

In research group, we also have measured mutual diffusion for multicomponent systems, that is, for ternary and more recently quaternary systems.

Diffusion in a ternary solution is described by the diffusion equations (Equations (3) and (4)).

$$-(J_1) = (D_{11})_v \frac{\partial c_1}{\partial x} + (D_{12})_v \frac{\partial c_2}{\partial x} \tag{3}$$

$$-(J_2) = (D_{21})_v \frac{\partial c_1}{\partial x} + (D_{22})_v \frac{\partial c_2}{\partial x} \tag{4}$$

where J_1, J_2, $\frac{\partial c_1}{\partial x}$, and $\frac{\partial c_2}{\partial x}$ are the molar fluxes and the gradients in the concentrations of solute 1 and 2, respectively. The index v represents the volume fixed frame of the reference used in these measurements. Main diffusion coefficients give the flux of each solute produced by its own concentration gradient. Cross diffusion coefficients D_{12} and D_{21} give the coupled flux of each solute driven by the concentration gradient in the other solute. A positive D_{ik} cross coefficient ($i \neq k$) indicates co-current coupled transport of solute i from regions of higher to lower concentrations of solute k. However, a negative D_{ik} coefficient indicates counter current coupled transport of solute i from regions of lower to higher concentration of solute k.

Recently, diffusion in a quaternary solution has been described by the diffusion equations[1] (Equations 5–7) [19],

$$-(J_1) = {}^{123}(D_{11})_v \frac{\partial c_3}{\partial x} + {}^{123}(D_{12})_v \frac{\partial c_2}{\partial x} + {}^{123}(D_{13})_v \frac{\partial c_3}{\partial x} \tag{5}$$

$$-(J_2) = {}^{123}(D_{21})_v \frac{\partial c_1}{\partial x} + {}^{123}(D_{22})_v \frac{\partial c_2}{\partial x} + {}^{123}(D_{23})_v \frac{\partial c_3}{\partial x} \tag{6}$$

---

[1]An aqueous quaternary system, which for brevity we will indicate with ijk, not indicating the solvent 0, has three corresponding aqueous ternary systems (ij, ik, and jk), and three corresponding aqueous binary systems (i, j, and k). The main term quaternary diffusion coefficients can then be compared with two ternary values, ${}^{ij}D_{ii}$ and ${}^{ik}D_{ii}$, and with one binary value; similarly for the other two main terms ${}^{ijk}D_{jj}$ and ${}^{ijk}D_{kk}$. The quaternary cross diffusion coefficient ${}^{ijk}D$ij can be compared only with one ternary diffusion coefficient ${}^{ij}D_{ij}$; this is also true for all the other cross terms. The comparison between the diffusion coefficients of system ijk with those of the systems ij, ik, and jk, permits to obtain information on the effect of adding each solute to the other two. The comparison between the diffusion coefficients of the quaternary system with those of the systems ijk, and with those of the systems i, j, and k, permits to obtain information on the effect of adding each couple of solutes to the other one.

$$-(J_3) = {}^{123}(D_{31})_v \frac{\partial c_1}{\partial x} + {}^{123}(D_{32})_v \frac{\partial c_2}{\partial x} + {}^{123}(D_{33})_v \frac{\partial c_3}{\partial x} \quad (7)$$

where the main diffusion coefficients, ${}^{123}D_{11}$, ${}^{123}D_{22}$, and ${}^{123}D_{33}$, give the flux of each solute produced by its own concentration gradient. Cross diffusion coefficients ${}^{123}D_{12}$, ${}^{123}D_{13}$, ${}^{123}D_{21}$, ${}^{123}D_{23}$, ${}^{123}D_{31}$, and ${}^{123}D_{32}$ give the coupled flux of each solute driven by a concentration gradient in the other solute.

2.2 EXPERIMENTAL TECHNIQUES: CONDUCTIMETRIC AND TAYLOR DISPERSION TECHNIQUES

Experimental methods that can be employed to determine mutual diffusion coefficients [20, 21]: Diaphragm Cell (inaccuracy 0.5–1%), Conductometric (inaccuracy 0.2%), Gouy and Rayleigh Interferometry (inaccuracy< 0.1%), and Taylor Dispersion (inaccuracy 1–2%). While the first and second methods consume days in experimental time, the last ones imply just hours. The conductometric technique follows the diffusion process by measuring the ratio of electrical resistances of the electrolyte solution in two vertically opposed capillaries as time proceeds. Despite this method has given us reasonably precise and accurate results, it is limited to studies of mutual diffusion in electrolyte solutions, and like in diaphragm cell experiments, the run times are inconveniently long (~days). The Gouy Method also has high precision, but when applied to microemulsions they are prone to gravitational instabilities and convections. Thus, the Taylor dispersion has become increasingly popular for measuring diffusion in solutions, because of its experimental short time and its major application to the different systems (electrolytes or non-electrolytes). In addition, with this method it is possible to measure multicomponent mutual diffusion coefficients.

Mutual differential diffusion coefficients of binary (e.g. [6-8]) and pseudo binary systems (such as, e.g., cobalt chloride in aqueous solutions of sucrose [9]), have been measured using a conductometric cell and an automatic apparatus to follow diffusion. This cell uses an open-ended capillary method and a conductometric technique is used to follow the diffusion process by measuring the resistance of the solution inside the capillaries, at recorded times. Figure 1 shows a schematic representation of the open-ended capillary cell.

The theory of the Taylor dispersion technique is well described in the literature [11-20], and so the authors only indicate some relevant points concerning this method on the experimental determination of binary and ternary diffusion coefficients (Figure 2).

It is based on the dispersion of small amounts of solution injected into laminar carrier streams of solvent or solution of different composition, flowing through a long capillary tube. The length of the Teflon dispersion tube used in the present study was measured directly by stretching the tube in a large hall and using two high quality theodolites and appropriate mirrors to accurately focus on the tube ends. This technique gave a tube length of 3.2799 (± 0.0001) × 10^4 mm, in agreement with less precise control measurements using a good quality measuring tape. The radius of the tube, 0.5570 (± 0.0003) mm, was calculated from the tube volume obtained by accurately weighing (resolution 0.1 mg) the tube when empty and when filled with distilled water of known density. At the start of each run, a 6-port Teflon injection valve (Rheodyne, model 5020) was used to introduce 0.063 ml of solution into the laminar carrier stream of slightly different composition. A flow rate of 0.17 ml min^{-1} was maintained by a metering pump (Gilson model Minipuls 3) to give retention times of about 1.1×10^4 s. The dispersion tube and the injection valve were kept at 298.15K and 310.15K (± 0.01K) in an air thermostat.

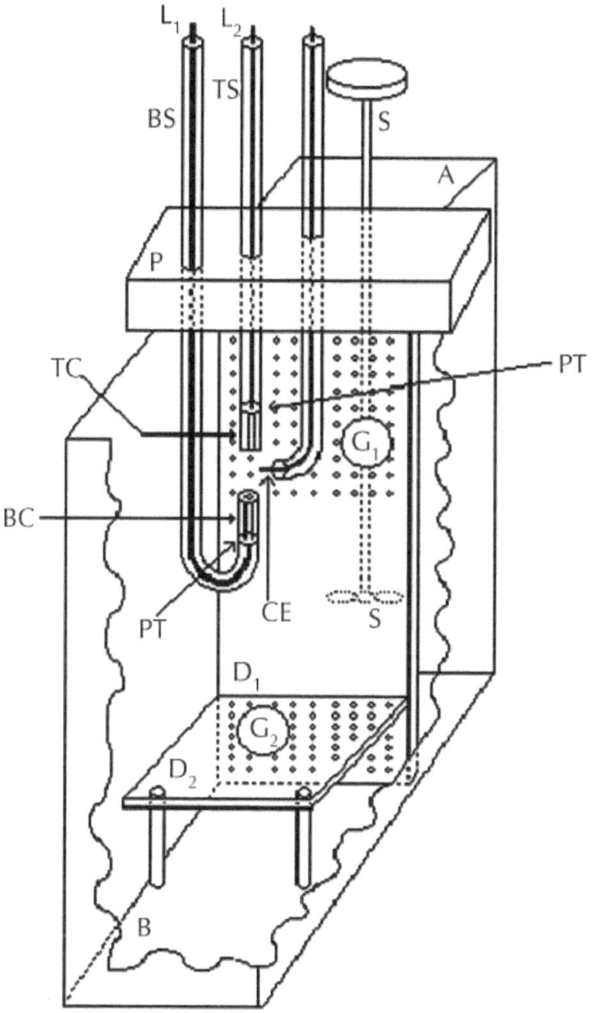

FIGURE 1 The TS and BS: support capillaries, TC and BC: top and bottom diffusion capillaries, CE: central electrode, PT: platinum electrodes, D_1 and D_2: perspex sheets, S: glass stirrer, P: perspex block, G_1 and G_2: perforations in perspex sheets, A and B: sections of the tank, and L_1 and L_2: small diameter coaxial leads [1].

Dispersion of the injected samples has been monitored using a differential refractometer (Waters model 2410) at the outlet of the dispersion tube. Detector voltages, $V(t)$, were measured at accurately 5 s intervals with a digital voltmeter (Agilent 34401A) with an IEEE interface.

Binary diffusion coefficients have been evaluated by fitting the dispersion Equation (8) to the detector voltages.

$$V(V(t) = V_0 + V_1 t + V_{max} (t_R/t)^{1/2} \exp[-12D(t - t_R)^2/r^2 t] \tag{8}$$

where r is the internal radius of Teflon dispersion tube. The additional fitting parameters were the mean sample retention time t_R, peak height V_{max}, baseline voltage V_0, and baseline slope V_1. Gravitational instabilities and unwanted convection are negligible because the carrier is confined to narrow-bore capillary tubing.

Extensions of the Taylor technique have been used to measure ternary mutual diffusion coefficients (D_{ik}) for multicomponent solutions. These D_{ik} coefficients, defined by Equations (3) and (4), were evaluated by fitting the ternary dispersion equation (Equation (9)) to two or more replicate pairs of peaks for each carrier stream.

$$V(t) = V_0 + V_1 t + V_{max} (t_R/t)^{1/2} \left[W_1 \exp\left(-\frac{12D_1 (t-t_R)^2}{r^2 t} \right) + (1-W_1) \exp\left(-\frac{12D_2 (t-t_R)^2}{r^2 t} \right) \right] \tag{9}$$

Two pairs of refractive index profiles, D_1 and D_2, are the eigenvalues of the matrix of the ternary D_{ik} coefficients. In these experiments, small volumes of ΔV of solution, of composition $\bar{c}_1 + \overline{\Delta c_1}$ and $\bar{c}_2 + \overline{\Delta c_2}$ are injected into carrier solutions of composition $\bar{c}_1$ and $\bar{c}_2$, at time t = 0.

Extensions of the Taylor technique have been used to measure quaternary mutual diffusion coefficients $^{ijk}(D_{ij})$ for multicomponent solutions. These $^{ijk}(D_{ij})$ coefficients, defined by Equation (5–7), were evaluated by fitting the quaternary dispersion equation (Eqution (10)).

$$\mathrm{V(t) = V_0 + V_1 t + K} \sum_{i=1}^{3} R_i \left[c_i(\mathrm{t}) - \bar{c}_i \right] \tag{10}$$

where $K = dV/dn$ is the sensitivity of the detector, being n the refractive index, $R_i = dn/d\,\bar{c}_i$ measures the change in the detected property per unit change in the concentration of solute, and $c_i(\mathrm{t}) - \bar{c}_i$ represents the dispersion solute average concentrations given by:

$$c_i(\mathrm{t}) = \bar{c}_i + \frac{2\Delta v}{\pi r^3 u} \left(\frac{3}{\pi t} \right)^{1/2} \sum_{k=1}^{3} \sum_{p=1}^{3} A_{ik} B_{kp} \Delta C_p D_k^{1/2} \exp\left[-12D_k (t-t_R)^2 / r^2 t \right] \tag{11}$$

The D_k are the eigenvalues of the matrix D of quaternary diffusion coefficients. The columns of matrix A are independent eigenvectors of D and B is the inverse of A.

The terms $V_0 + V_1$t are often included in practice to allow for small drifts in the detector signal. In these experiments, small volumes of ΔV of solution, of composition

$\bar{c}_1+\overline{\Delta c}_1$, $\bar{c}_2+\overline{\Delta c}_2$, and $\bar{c}_3+\overline{\Delta c}_3$ are injected into carrier solutions of composition, $\bar{c}_1$, $\bar{c}_2$, and $\bar{c}_3$ at time t = 0.

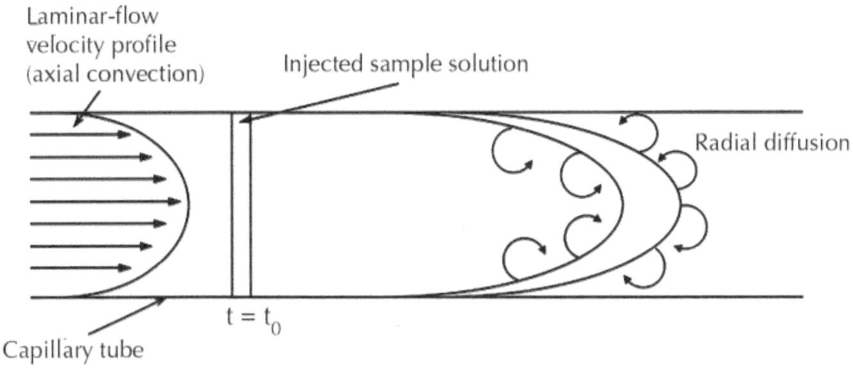

FIGURE 2 Schematic representation of the Taylor dispersion technique [22].

2.3 SOME EXPERIMENTAL DISCUSSION AND RESULTS

Mutual differential diffusion coefficients of several electrolytes 1:1, 2:2, and 2:1, in different media (considering these systems as binary or pseudo binary systems, depending on the circumstances) have been measured using a conductometric cell [1]. The already published mutual differential diffusion coefficients data are average results of, at least, three independent measurements. The imprecision of such average results is, with few exceptions, lower than 1%.

The calculation of diffusion coefficients from equations based on some models describing the movement of matter in electrolyte solutions, in the end, a process contributing to the knowledge of their structure, provided we have accurate experimental data to test these equations. Thus, to understand the behavior of transport process of these aqueous systems, experimental mutual diffusion coefficients have been compared with those estimated using several equations, resulting from different models.

Assuming that each ion of the diffusing electrolyte can be regarded as moving under the influence of two forces: (i) a gradient of the chemical potential for that ionic species, and (ii) an electrical field produced by the motion of oppositely charged ions, we come up to the Nernst-Hartley equation [20, 21].

$$D = [(\nu_1 + \nu_2)\,\lambda_1^{\,0}\lambda_2^{\,0}/(\nu_1\,|Z_1|\,(\lambda_1^{\,0} + \lambda_2^{\,0}))]\,(R\,T/F^2)\,[1 + (d\ln\gamma_\pm/d\ln c)] \qquad (12)$$

where λ^0 are the limiting ionic conductivities of the ions (subscripts 1 and 2 for cation and anion, respectively), Z is the algebraic valence of the ion, ν is the number of ions formed upon complete ionization of one solute "molecule", T is the absolute temperature, R and F are the gas and Faraday constants, respectively, and $\gamma_\pm$ is the mean molar activity coefficient.

Equation (12) is often written as:

$$D = D^0\,[1 + (d\ln\gamma_\pm/d\ln c)] \qquad (13)$$

where D^0 is the Nernst limiting value of the diffusion coefficient.

Onsager and Fuoss [23] improved Equation (13) by taking into account the electrophoretic effects (Equation 14):

$$D = D^0 [1 + (d \ln\gamma_{\pm}/d \ln c)] \tag{14}$$

The difference between Equations (13) and (14) can be found in the electrophoretic term, Δ_n, given by:

$$\Delta_n = K_B T A_n (Z_1{}^n t_2{}^0 + Z_2{}^n t_1{}^0)^2/(a^n |Z_1 Z_2|)$$

where K_B is the Boltzmann's constant, A_n is functions of the dielectric constant, viscosity of the solvent, temperature, and dimensionless concentration-dependent quantity (κ, a), being κ the reciprocal of average radius of the ionic atmosphere, and $t_1{}^0$ and $t_2{}^0$ are the limiting transport numbers of the cation and anion, respectively.

Since the expression for the electrophoretic effect has been derived on the basis of the expansion of the Boltzmann exponential function, because that function had been consistent with the Poisson equation, we only, in major cases, would have to take into account the electrophoretic term of the first order (n = 1). For symmetrical electrolytes we can consider the second term. Thus, the experimental data can be compared with the calculated D on the basis of Equations (16) and (17) for symmetrical and non-symmetrical electrolytes, respectively.

$$D = (D^0 + \Delta_1 + \Delta_2) [1 + c (d \ln\gamma_{\pm}/d c)] \tag{16}$$

$$D = (D^0 + \Delta_1) [1 + c (d \ln\gamma_{\pm}/d c)] \tag{17}$$

The theory of mutual diffusion in binary electrolytes, developed by Pikal [24], includes the Onsager-Fuoss equation, but has new terms resulting from the application of the Boltzmann exponential function for the study of diffusion. The eventual formation of ion pairs is taken into account in this model, not considered in the Onsager-Fuoss'.

$$D = \frac{1}{\frac{1}{M^0}(1 - \frac{\Delta M}{M^0})} (10^3 R T \nu) [1 + c (d \ln\gamma_{\pm}/d c)] \tag{18}$$

Data from these models for different types of electrolytes in dilute aqueous solutions have been presented in the literature [25, 26]. From those data we conclude that for symmetrical uni-univalent, both theories (Onsager and Pikal) give similar results, and they are consistent with experimental ones. In fact, if Pikal's theory is valid, ΔM^{0F} must be the major term; all other terms are much smaller and they partially cancel each other. Concerning symmetrical but polyvalent electrolytes [25, 26], we can well see that Pikal's theory is a better approximation than the Onsager-Fuoss'. The ion association, taken into account in this model [27], can justify this behavior.

In polyvalent non-symmetrical electrolytes, agreement between experimental data and Pikal calculations is not so good, eventually because of the full use of Boltzmann's exponential in Pikal's development.

Although no theory on diffusion in electrolyte solutions is capable of giving generally reliable data on D, we suggest, for estimating purposes, when no experimental data are available, the calculations of D_{OF} and D_{Pikal} for hundreds of electrolytes already made by Lobo et al. [25, 26]. That is, for symmetrical uni-valent electrolyte (1:1) we suggest the application of Onsager-Fuoss equation with any a (ion size) from the literature (e.g., Lobo's publication), because parameter a has little effect on final conclusions of D_{OF}; for symmetrical polyvalent (basically 2:2), we suggest the application of Pikal equation. In this case, because D_{Pikal} is strongly affected by the choice of a, we suggest calculation with two (or more) reasonable values of a, assuming that the actual value of D should lie between them, for non-symmetrical polyvalent, we suggest both Onsager-Fuoss and Pikal theories, assuming the actual value of D should lie between them. Now, the choice of a is irrelevant, within reasonable limits.

Concerning more concentrated solutions, no definite conclusion is possible. In fact, the results predicted from these models differ from experimental observation (ca >4%). This is not surprising if we take into account the change with concentration of parameters such as viscosity, dielectric constant, hydration, and hydrolysis which are not taken into account in these models [20, 21, 28].

For example, the experimental diffusion coefficient values of $CrCl_3$ in dilute solutions at 298.15K [6] are higher than the calculated ones (D_{OF}).This can be explained not only by the initial $CrCl_3$ gradient and the formation of complexes between chloride and chromium (III), but also by a further hydrogen ion flux, according to Equation (19).

$$x\, Cr^{3+} + 2y\, H_2O \leftrightarrows Cr_x(OH)_y^{(3x-y)+} + y\, H_3O^+ \quad (19)$$

Similar situations have been found for the systems $BeSO_4$/water [3], $CoCl_2$ chloride/water [4], and $Pb(NO_3)_2$/water [5].

In fact, comparing the estimated diffusion coefficients of $Pb(NO_3)_2$, D_{OF}, with the related experimental values [5], an increase in the experimental D values is found in lead (II) nitrate concentrations below 0.025M. This can be explained not only by the initial $Pb(NO_3)_2$ gradient, but also by a further H_3O^+ flux. Consequently, as H_3O^+ diffuses more rapidly than NO_3^- or Pb^{2+}, the lead(II) nitrate gradient generates "its own" HNO_3 flux. Thus, the $Pb(NO_3)_2$/water mixture should be considered a ternary system. However, in the present experimental conditions we may consider the system as pseudo-binary, mainly for c ≥0.01M, and consequently, take the measured parameter, D_{av}, as the main diffusion coefficient, D_{11}.

For c <0.01M, we can estimate the concentration of H_3O^+ produced by hydrolysis of Pb(II) using Equations (20) and (21), assuming that: (a) the fluxes of the species, HNO_3 and $Pb(NO_3)_2$, are independent, (b) the values of the diffusion coefficients, D_{OF}, come from Equation. (3). The percentages of H_3O^+ (or the amount of acid that would be necessary to add to one solution of $Pb(NO_3)_2$ in the absence of hydrolysis, resulting in this way a simulation of a more real system) are estimated from the following equations [5]:

$$\alpha\, D_{OF}HNO_3) + \beta\, D_{OF}Pb(NO_3)_2) = D_{av} \tag{20}$$

$$\alpha + \beta = 1 \tag{21}$$

where $\alpha \times 100$ and $\beta \times 100$ are the percentages of nitric acid and lead nitrate, respectively. From those data, we can conclude that, for c $\geq$0.01M, α becomes very low, suggesting that either the hydrolysis effect or the contribution of $D_{of}(Pb(NO_3)_2)$ to the whole diffusion process, can be neglected. Tables (1–3) give the estimated percentage of hydrogen ions, α, resulting from the hydrolysis of Pb^{2+}, Be^{2+}, and Co^{2+} ions in aqueous solutions of lead (II) nitrate, beryllium sulfate, and cobalt chloride, respectively.

TABLE 1 Estimated percentage of hydrogen ions, α, resulting from the hydrolysis of Pb^{2+} in aqueous solutions of lead (II) nitrate at 298.15K, using Equations (20) and (21) [5].

$[Pb(NO_3)_2]/(mol\ dm^{-3})$	α/%
0.001	26.0
0.005	7.2
0.01	2.5
0.05	a)

a) For this concentration we can consider α as non-relevant.

It what concerns binary systems involving non-electrolytes, we have been measuring mutual diffusion coefficients of some cyclodextrins (α-CD, β-CD, HP-α-CD, and HP-β-CD) [12, 13] and some drugs (e.g., caffeine and isoniazid [14]) in aqueous solutions. Also, from comparison of these experimental diffusion coefficients with the related calculated values, it is possible to give some structural information, such as diffusion coefficients at infinitesimal concentration at different temperatures, estimation of activity coefficients by using equations of Hartley and Gordon, estimation of hydrodynamic radius, and estimation of activation energies, *Ea*, of the diffusion process at several temperatures.

TABLE 2 Estimated percentage of hydrogen ions, α, resulting from the hydrolysis of Be^{2+} in aqueous solutions of beryllium sulfate at 298.15K, using similar equation to the Equations (20) and (21) [3].

$[BeSO_4]/(mol\ dm^{-3})$	α/%
0.003	**69.0**
0.005	**48.0**
0.008	**26.0**
0.010	**24.0**

TABLE 3 Estimated percentage of hydrogen ions, α, resulting from the hydrolysis of Co^{2+} in aqueous solutions of cobalt (II) chloride at 298.15K, using equation similar to Equations (20) and (21) [4].

[$CoCl_2$/(mol dm^{-3})	α/%
0.001	6.0
0.003	6.0
0.005	6.0
0.008	8.0
0.010	5.0

Also, from the measurements of diffusion coefficients of the ternary systems already studied (e.g., β-cyclodextrin plus caffeine [15], 2-hydroxypropyl-β-cyclodextrin plus caffeine [16], $CuCl_2$ (1) plus caffeine [10], and KCl plus theophylline (THP) [18]), it is possible to give a contribution to the understanding of the structure of electrolyte solutions and their thermodynamic behavior. For example, by using Equations (22) and (23), and through the experimental tracer ternary diffusion coefficients of KCl dissolved in supporting THP solutions, D_{11}^0 $(c_1/c_2 = 0)$ and tracer ternary diffusion coefficients of THP dissolved in supporting KCl solutions, D_{22}^0 $(c_2/c_1 = 0)$ [18], it will be possible to estimate some parameters, such as the diffusion coefficient of the aggregate between KCl and THP [18] and the respective association constant.

$$D_{11}^0(c_1/c_2 = 0) = 2D_{Cl^-} \frac{(X_1)\, D^0_{\text{complex}^+} + (1-X_1)\; D^0_{K^+}}{D_{Cl^-} + (X_1)\, D^0_{\text{complex}^+} + (1-X_1)\; D^0_{K^+}} \tag{22}$$

$$D_{22}^0(c_2/c_1 = 0) = X_2\, D^0_{\text{complex}} + (1-X_2)\, D^0_{\text{THP}} \tag{23}$$

2.4 CONCLUSION

The study of diffusion processes of electrolytes and non-electrolytes in aqueous solutions is important for fundamental reasons, helping to understand the nature of aqueous electrolyte structure, for practical applications in fields such as corrosion, and provide transport data necessary to model diffusion in pharmaceutical applications. Although no theory on diffusion in electrolyte or non-electrolyte solutions is capable of giving generally reliable data on D, there are, however, estimating purposes, whose data, when compared with the experimental values, will allow us to take off conclusions on the nature of the system.

KEYWORDS

- **Aqueous solutions**
- **Diffusion coefficients**
- **Electrolytes**
- **Nonionic solutes**
- **Transport properties**

REFERENCES

1. Agar, J. N. and Lobo, V. M. M. *J. Chem. Soc., Faraday Trans. I*, **71**, 1659–1666 (1975).
2. Lobo, V. M. M. *Handbook of Electrolyte Solutions*. Elsevier, Amsterdam (1990).
3. Lobo, V. M. M., Ribeiro, A. C. F., and Veríssimo, L. P. *J. Chem. Eng. Data*, **39**, 726–728 (1994).
4. Ribeiro, A. C. F., Lobo, V. M. M., and Natividade, J. J. S. *J. Chem. Eng. Data*, **47**, 539–541 (2002).
5. Valente, A. J. M., Ribeiro, A. C. F., Lobo, V. M. M., and Jiménez, A. *J. Mol. Liq.*, **111**, 33–38 (2004).
6. Ribeiro, A. C. F., Lobo, V. M. M., Oliveira, L. R. C., Burrows, H. D., Azevedo, E. F. G., Fangaia, S. I. G., Nicolau, P. M. G., and Guerra, F. A. D. R. A. *J. Chem. Eng. Data*, **50**, 1014–1017 (2005).
7. Ribeiro, A. C. F., Lobo, V. M. M., Valente, A. J. M., Simões, S. M. N., Sobral, A. J. F. N., Ramos, M. L., and Burrows, H. D. *Polyhedron*, **25**, 3581–3587 (2006).
8. Ribeiro, A. C. F., Esteso M. A., Lobo, V. M. M., Valente, A. J. M., Sobral, A. J. F. N., and Burrows, H. D. *Electrochim. Acta*, **52**, 6450–6455 (2007).
9. Ribeiro, A. C. F., Costa, D. O., Simões, S. M. N., Pereira, R. F. P., Lobo, V. M. M., Valente, A. J. M., and Esteso, M. A. *Electrochim. Acta*, **55**, 4483–4487 (2010).
10. Ribeiro, A. C. F., Esteso, M. A., Lobo, V. M. M., Valente, A. J. M., Simões, S. M. N., Sobral, A. J. F. N., Ramos, L., Burrows, H. D., Amado, A. M., and Amorim da Costa, A. M. *J. Carbohydr. Chem*, **25**, 173–185 (2006).
11. Ribeiro, A. C. F., Lobo, V. M. M., Leaist, D. G., Natividade, J. J. S., Veríssimo, L. P., Barros, M. C. F., Cabral, A. M. T. D. P. V. *J. Sol. Chem*, **34**, 1009–1016 (2005).
12. Ribeiro, A. C. F., Leaist, D. G., Esteso, M. A., Lobo, V. M. M., Valente, A. J. M., Santos, C. I. A. V., Cabral, A. M. T. D. P. V., and Veiga, F. J. B. *J. Chem. Eng. Data*, **51**, 1368–1371 (2006).
13. Ribeiro, A. C. F., Valente, A. J. M., Santos, C. I. A. V., Prazeres, P. M. R. A., Lobo, V. M. M., Burrows, H. D., Esteso, M. A., Cabral, A. M. T. D. P. V., and Veiga, F. J. B. *J. Chem. Eng. Data*, **52**, 586–590 (2007).
14. Ribeiro, A. C. F., Santos, A. C. G., Lobo, V. M. M., Veiga, F. J. B., Cabral, A. M. T. D. P. V., Esteso, M. A., and Ortona, O. *J. Chem. Eng Data*, **54**, 3235–3237 (2009).
15. Ribeiro, A. C. F., Santos, C. I. A. V., Lobo, V. M. M., Cabral, A. M. T. D. P. V., Veiga, F. J. B., and Esteso, M. A. *J. Chem. Eng. Data*, **54**, 115–117 (2009).
16. Ribeiro, A. C. F., Santos, C. I. A. V., Lobo, V. M. M., Cabral, A. M. T. D. P. V., Veiga, F. J. B., and Esteso, M. A. *J. Chem. Thermodynamics*, **41**, 1324–1328 (2009).
17. Ribeiro, A. C. F., Simões, S. M. N., Lobo, V. M. M., Valente, A. J. M., and Esteso, M. A. *Food Chem.*, **118**, 847–850 (2010).
18. Santos, C. I. A. V., Lobo, V. M. M., Esteso, M. A., Ribeiro, A. C. F. *Effect of Potassium Chloride on Diffusion of Theophylline at T = 298.15 K*. In press.
19. Ribeiro, A. C. F., Santos, C. I. A. V., Lobo, V. M. M., and Esteso, M. A. *J. Chem. Eng Data*, **55**, 2610–2612 (2010).
20. Robinson, R. A. and Stokes, R. H. *Electrolyte Solutions*, 2nd Ed. Butterworths, London (1959).
21. Tyrrell, H. J. V. and Harris, K. R. *Diffusion in Liquids*, 2nd Ed. Butterworths, London (1984).
22. Callendar, R. and Leaist, D. G. *J. Sol. Chem.*, **35**, 353–379 (2006).

23. Onsager, L. and Fuoss, R. M. *J. Phys. Chem.*, **36**, 2689–2778 (1932).
24. Pikal, M. J. *J. Phys. Chem.*, **75**, 663–680 (1971).
25. Lobo, V. M. M., Ribeiro, A. C. F., and Andrade, S. G. C. S. *Ber. Buns. Phys. Chem.*, **99**, 713–720 (1995).
26. Lobo, V. M. M., Ribeiro, A. C. F., and Andrade, S. G. C. S. *Port. Electrochim. Acta*, **14**, 45–124 (1996).
27. Lobo, V. M. M. and Ribeiro, A. C. F. *Port. Electrochim. Acta*, **12**, 29–41 (1994).
28. Lobo, V. M. M. and Ribeiro, A. C. F. *Port. Electrochim. Acta*, **13**, 41–62 (1995).

3 Frictional and Elastic Components of the Viscosity of Concentrated Solutions Polystyrene in Toluene

Yu. G. Medvedevskikh, O. Yu. Khavunko, L. I. Bazylyak, and G. E. Zaikov

CONTENTS

3.1 INTRODUCTION

The viscosity (η) of polymeric solutions is an object of the numerous experimental and theoretical investigations generalized by Ferry [1], De Jennes [2], Tsvietkov et al. [3], and Malkin and Isayev [4]. This is explained both by the practical importance of the presented property of polymeric solutions in a number of the technological processes and by the variety of the factors having an influence on the η value, also by a wide diapason (from 10^{-3} to 10^{2} $Pa \times s$) of the viscosity change under transition from the diluted solutions and melts to the concentrated ones. It gives a great informational groundwork for the testing of different theoretical imaginations about the equilibrium and dynamic properties of the polymeric chains.

It can be marked three main peculiarities for the characteristic of the concentrated polymeric solutions viscosity, namely:

(1) Measurable effective viscosity η for the concentrated solutions is considerable stronger than η for the diluted solutions and depends on the velocity gradient g of the hydrodynamic flow or on the shear rate. Malkin and Isayev [4] distinguish the initial η_0 and the final η_∞ viscosities ($\eta_0 > \eta_\infty$), to which the extreme conditions $g \to 0$ and $g \to \infty$ correspond respectively.

(2) Strong power dependence of η on the length N of a polymeric chain and on the concentration ρ (g/m³) of a polymer into solution exists: $\eta \sim \rho^{\alpha} N^{\beta}$ with the indexes $\alpha = 5 \div 7$, $\beta = 3{,}3 \div 3{,}5$, as it was shown by Malkin and Isayev [4].

(3) It was experimentally determined by Ferry and Grassley [5] that the viscosity η and the characteristic relaxation time t^* of the polymeric chains into concentrated solutions are characterized by the same scaling dependence on the length of a chain:

$$\eta \sim \rho^{\alpha} N^{\beta} \tag{1}$$

with the index β = 3, 4.

Among the numerous theoretical approaches to the analysis of the polymeric solutions viscosity anomaly, that is the dependence of η on g, it can be marked the three main approaches. The first one connects the anomaly of the viscosity with the influence of the shear strain on the potential energy of the molecular kinetic units transition from the one equilibrium state into another one and gives the analysis of this transition from the point of view of the absolute reactions rates theory (prior work by Glesston [6]). It was proposed, for example, the equation:

$$\eta = A \cdot f \exp\{(a - bf)/kT\} \tag{2}$$

which predicts the viscosity decreasing at the increase of the shear force f at A, a and b, which are some constants.

However, the Equation (2) has not take into account the specificity of the polymeric chain, that is why, it was not win recognized in the viscosity theory of the polymeric solutions.

In accordance with the second approach the polymeric solutions viscosity anomaly is explained by the effect of the hydrodynamic interaction between the links of the polymeric chain, such links represent by themselves the "beads" into the "necklace" model. Accordingly to this effect the hydrodynamic flow around the presented "bead" essentially depends on the position of the other "beads" into the polymeric ball. An anomaly of the viscosity was conditioned by the anisotropy of the hydrodynamic interaction which creates the orientational effect (prior work by Peterlin and Copic [7]). High values of the viscosity for the concentrated solutions and its strong gradient dependence cannot be explained only by the effect of the hydrodynamic interaction.

That is why, the approaches integrated into the conception of the structural theory of the viscosity were generally recognized. In accordance with this theory the viscosity of the concentrated polymeric solutions is determined by the quasinet of the linkages of twisted between themselves polymeric chains and, therefore, depends on the modulus of elasticity E of the quasinet and on the characteristic relaxation time t^* (original works by Ferry [1] and by De Jennes [2]):

$$\eta = E \cdot t^* \tag{3}$$

It is supposed that the E is directly proportional to the density of the linkages assemblies and is inversely proportional to the interval between them along the same chain. An anomaly of the viscosity is explained by the linkages assemblies' density decreasing at their destruction under the action of shear strain (prior work by Hoffman and Rother [8]), by the change of the relaxation spectrum (prior work by Leonov and Vinogradov [9]), and the distortion of the polymer chain links distribution function relatively to its center of gravity (prior work by Williams [10]). A gradient dependence of the viscosity is described by the equation of Williams [10]:

$$(\eta - \eta_\infty)/(\eta_0 - \eta_\infty) = f\left(gt^*\right) \tag{4}$$

It was greatly recognized the universal scaling ratio by Ferry [1] and by Grassley [5]:

$$\eta = \eta_0 \cdot f\left(gt^*\right) \tag{5}$$

in which the dimensionless function $f\left(gt^*\right) = f(x)$ has the asymptotes $f(0) = 1$, $f(x)_{x>>1} = x^{-\gamma}$, $\gamma = 0, 8$.

Hence, both Equations (4) and (5) declare the gradient dependence of η by the function of the one non–dimensional parameter gt^*. However, under the theoretical estimation of η and t^* as a function of N there are contradictions between the experimentally determined ratio Equation (1) and $\beta = 3, 4$. Thus, the analysis of the entrainment of the surrounding chains under the movement of some separated chain by Bueche [11] leads to the dependencies $\eta \sim N^{3,5}$ but $t^* \sim N^{4,5}$. At the Edwards and Grant [12] analysis of the self-coordinated movement of a chain enclosing into the tube formed by the neighboring chains it was obtained the $\eta \sim N^3$, $t^* \sim N^4$. The approach of De Jennes [13] based on the conception of the reptational mechanism of the polymeric chain movement gives the following dependence $\eta \sim t^* N^3$. So, the index $\beta = 3, 4$ in the ratio (1) from the point of view of Ferry [1] remains by one among the main unsolved tasks of the Polymers Physics.

Summarizing the presented short review, let us note that the conception about the viscosity elastic properties of the polymeric solutions accordingly to the Maxwell's equation should be signified the presence of two components of the effective viscosity, namely: the frictional one, caused by the friction forces only, and the elastic one, caused by the shear strain of the conformational volume of macromolecules. But in any among listed theoretical approaches the shear strain of the conformational volumes of macromolecules was not taken into account. The sustained opinion by Tsvietkov et al. [3] and by Malkin and Isayev [4] that the shear strain is visualized only in the strong hydrodynamic flows whereas it can be neglected at little g, facilitates to this fact. But in this case the inverse effect should be observed, namely an increase of η at the g enlargement.

This contradiction can be overpasses, if to take into account that although at the velocity gradient of hydrodynamic flow increasing the external action leading to the shear strain of the conformational volume of polymeric chain is increased, but at the same time, the characteristic time of the external action on the rotating polymeric ball is decreased, in accordance with the kinetic reasons this leads to the decreasing but not to the increasing of the shear strain degree. Such analysis done by Medvedevskikh [14] permitted to mark the frictional and the elastic components of the viscosity and to show that exactly the elastic component of the viscosity is the gradiently dependent value. The elastic properties of the conformational volume of polymeric chains, in particular shear modulus, were described early by Medvedevskikh [15, 16] based on the self–avoiding walks statistics (SAWS).

Here presented the experimental data concerning to the viscosity of the concentrated solutions of styrene in toluene and it is given their interpretation on the basis of works by Medvedevskikh [14–16].

It has been experimentally investigated the gradient dependence of the effective viscosity η for concentrated solutions of polystyrene in toluene at three concentrations $\rho = 0.4 \times 10^5$, $\rho = 0.5 \times 10^5$, and $\rho = 0.7 \times 10^5$ g/m^3 for the four fractions of polystyrene characterizing by the apparent molar weights $M = 5.1 \times 10^4$, $M = 4.1 \times 10^4$, $M = 3.3 \times 10^4$, and $M = 2.2 \times 10^4$ g/mol. The gradient dependence of a viscosity was studied at four temperatures, namely 25, 30, 35, and 40°C for each pair of ρ and M valuations.

The experiments have been carried out with the use of the RHEOTEST 2.1 viscosimeter at the different angular velocities ω (turns/s) of the working cylinder rotation. An analysis of the $\eta(\omega)$ dependencies permitted to mark out the frictional (η_f) and elastic (η_e) components of the viscosity and to study their dependence on temperature T, concentration ρ, and the length of a chain N.

It was determined that the main endowment into the frictional component of the viscosity has the relative movement of the twisted between themselves into m-ball polymeric chains. Such relative movement takes into account the all possible linkages effects. Elastic component of the viscosity η_e is determined by the elastic properties of the conformational volume of the m–ball of polymeric chains at its shear deformation.

It was experimentally confirmed the theoretical dependence $\eta_e \sim N^{3.4}\rho^{4.5}$ determined early, it leads to the well-known ratio $\eta_e \sim t_m^* \sim N^{3.4}$, which is, however, correct only for the elastic component of the viscosity.

On a basis of the experimental data it was obtained the numerical values of the characteristic time and of the activation energy for the segmental movement.

3.2 EXPERIMENTAL DATA AND STARTING POSITIONS

In order to obtain statistically significant experimental data we have studied the gradient dependence of the viscosity for the solution of polystyrene in toluene at concentrations 0.4×10^5, 0.5×10^5, and 0.7×10^5 g/m^3 for the four fractions of polystyrene characterizing by the apparent molar weights $M = 5.1 \times 10^4$, $M = 4.1 \times 10^4$, $M = 3.3$

$\times 10^4$, and $M = 2.2 \times 10^4$ g/mole. Molecular weights were determined by viscometry method with the use of the Oswald's viscometer having the capillary diameter 0.62×10^{-3} m. For each pair of values ρ and M the gradient dependence of the viscosity has been studied at fourth temperatures 25, 30, 35, and 40°C.

The experiments have been carried out with the use of the rotary viscometer *RHEOTEST* 2.1 with the working cylinder having two rotary surfaces with the diameters $d_1 = 3.4 \times 10^{-2}$ and $d_2 = 3.9 \times 10^{-2}$ m.

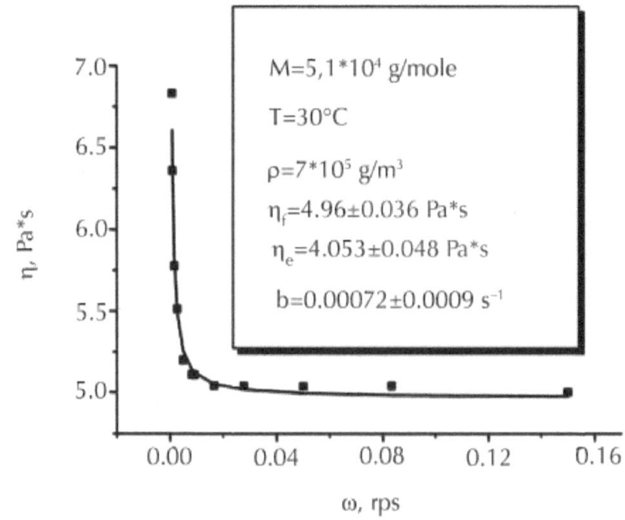

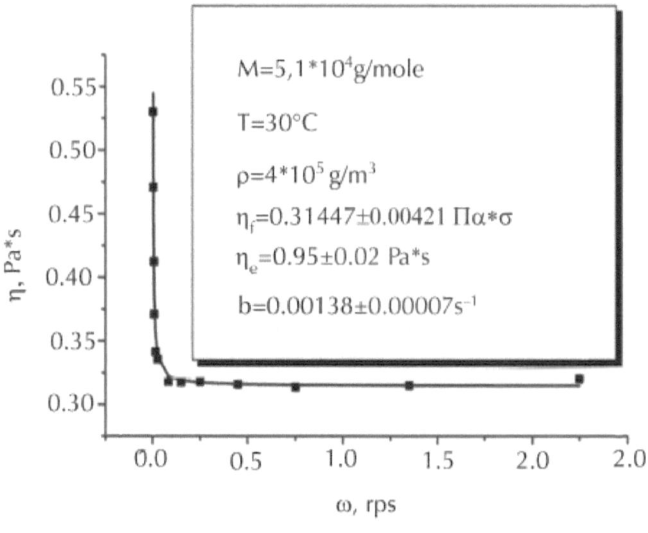

FIGURE 1 *(Continued)*

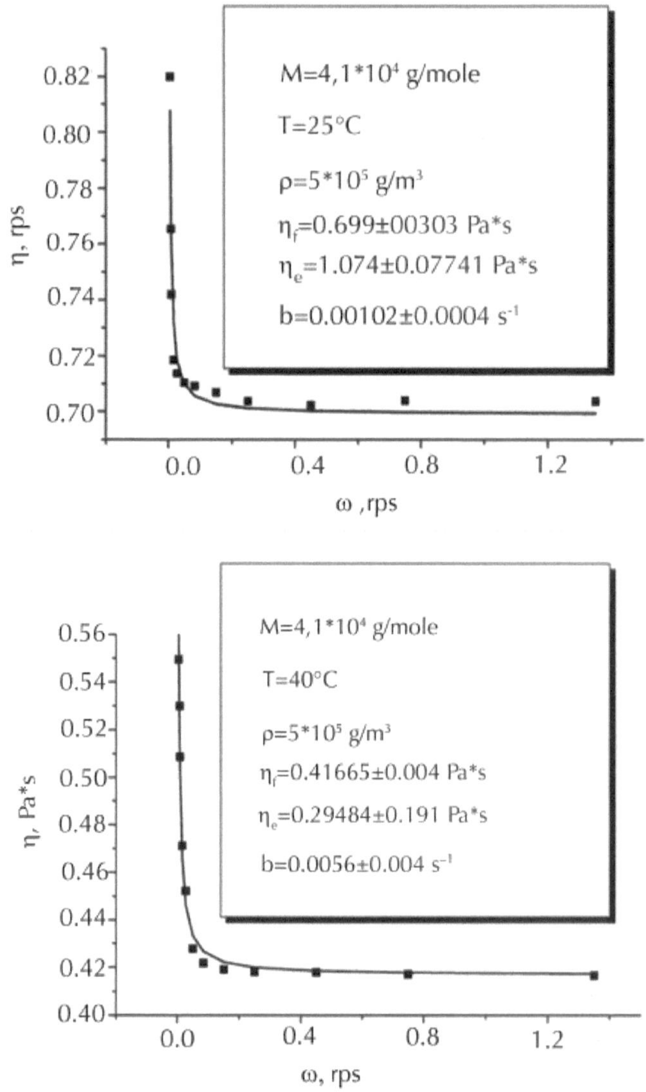

FIGURE 1 The typical experimental (points) and calculated according to the Equation (10) (full line graph) dependences of the effective viscosity η on the rotation rate of the working cylinder ω.

Typical dependencies of the viscosity η of solution on the angular velocity ω (turns/s) of the working cylinder rotation are represented on Figure 1. Generally it was obtained 48 curves of η*(*ω*)*.

For the analysis of the experimental curves of η*(*ω*)* it was used the equation by Medvedevskikh [14]:

$$\eta = \eta_f + \eta_e \left(1 - \exp\left\{-\frac{t_\nu^*}{t_m^*}\right\}\right) \Big/ \left(1 + \exp\left\{-\frac{t_\nu^*}{t_m^*}\right\}\right) \tag{6}$$

in which η is a measurable viscosity of the solution at given value of ω, η_f is the frictional and η_e is the elastic components of η, t_m^* is the characteristic time of the shear strain of the conformational volume of *m*–ball for the twisted polymeric chains, t_ν^* is the characteristic time of the hydrodynamic flow velocity gradient external action on the *m*–ball leading to its deformation and rotation .

The shear strain of the conformational volume of *m*–ball and its rotation is realized in accordance with the reptational mechanism presented by De Jennes [13], that is *via* the segmental movement of the polymeric chain, that is why t_m^* is also the characteristic time of the own, that is without the action *g*, rotation of *m*–ball (original works by Medvedevskikh [14]).

The Equation (6) leads to the two asymptotes:

$$\eta = \eta_f + \eta_e \qquad \text{at} \qquad t_\nu^* >> t_m^* \tag{7}$$

$$\eta = \eta_f \qquad \text{at} \qquad t_\nu^* << t_m^* \tag{8}$$

So, if the characteristic time t_ν^* of the external action *g* on the *m*–ball is considerably greater than the characteristic time t_m^* of its shear strain, than the effective viscosity is equal to a sum of the frictional and to the elastic components, whereas under condition of $t_\nu^* < t_m^*$ the shear strain of the *m*–ball due to the kinetic reasons has not time to be discovered and the measurable viscosity is equal only to the frictional component.

Characteristic time t_ν^* is a function of the velocity gradient of hydrodynamic flow which is formed by the rotation of the working cylinder into the rotary viscometer. So, it can be embed the ratio:

$$\frac{t_\nu^*}{t_m^*} = \frac{b}{\omega} \tag{9}$$

the substitution of which into (6) permits to rewrite it in the following form:

$$\eta = \eta_f + \eta_e \left(1 - \exp\left\{-\frac{b}{\omega}\right\}\right) \Big/ \left(1 + \exp\left\{-\frac{b}{\omega}\right\}\right) \tag{10}$$

The conditions $b/\omega >> 1$ and $b/\omega << 1$ in (10) corresponds to the conditions (7) and (8) in Equation (6).

In accordance with Equation (10), the effective viscosity *η(ω)* is a function on three parameters, namely η_f, η_e, and *b*. They can be found on a basis of the experimental values of *η(ω)* *via* the optimization method in program ORIGIN 5.0. As an analysis showed, the numerical values of η_f are easy determined upon a plateau on the curves η(ω) accordingly to the condition $b/\omega << 1$. However, the optimization method gave not always the correct values of η_e and *b*. There are two reasons for this. Firstly, in a field of the $\omega \to 0$ the uncertainty of η(ω) measurement is sharply increased since the moment of force registered by a device is a small. Secondly, in very important field of the curve transition η(ω) from the strong dependence of η on ω to the weak one the parameters η_e and *b* are interflowed into a composition $\eta_e b$, so they are by one parameter. Really, at the condition $b/\omega << 1$ decomposing the exponents into (10) and limiting by two terms of the row $\exp\left\{-\frac{b}{\omega}\right\} \approx 1 - \frac{b}{\omega}$, we will obtained $\eta = \eta_f + \eta_e b/2$. Due to the mentioned reasons the optimization method gives the values of η_e and *b* depending between them but does not giving the global minimum of the errors functional. That is why, at the estimation of η_e and *b* parameters it was necessary sometimes to supplement the optimization method with the "manual" method of the global minimum search varying mainly by the numerical estimation of η_e.

It is shown on Figure 1 that the calculated accordingly to Equation (10) curves of η(ω) and found in like manner parameters η_f, η_e, and *b* good describe the experimental data.

The results of the numerical estimations of η_f, η_e, and *b* for all 48 experimental curves η(ω) are represented in Table 1.

A review of these data shows that the all three parameters are the functions on the concentration of polymer into solution, on the length of a chain and on the temperature. But at this, the η_e and η_f are increased at the ρ and *M* increasing and are decreased at the *T* increasing, whereas, the *b* parameter is changed into the opposite way. The analysis of these dependencies will be represented. Here let us present the all needed for this analysis determinations, notifications, and information concerning to the concentrated polymeric solutions.

Investigated solutions of the polystyrene in toluene were concentrated, that is the following condition was performing for them:

$$\rho > \rho^* \tag{11}$$

Here ρ^* is a critical density of the solution per polymer corresponding to the starting of the polymeric chains conformational volumes overlapping having into diluted solution $(\rho \le \rho^*)$ the conformation of Flory ball by the radius:

$$R_f = aN^{3/5} \tag{12}$$

here *a* is a length of the chain's link.

It is followed from the determination of ρ^*:

$$\rho^* = \frac{M}{N_A R_f^3} = \frac{M_0 N}{N_A R_f^3} \tag{13}$$

TABLE 1 Optimization parameters η_f, η_e, and b in Equation (10).

$\rho \cdot 10^{-5}$, g/m^3		4,0				5,0				7,0			
T, °C	$M \cdot 10^4$ g/mole	5,1	4,1	3,3	2,2	5,1	4,1	3,3	2,2	5,1	4,1	3,3	2,2
	η_f, Pa·s	0,35	0,19	0,16	0,06	1,11	0,69	0,43	0,36	6,50	2,66	2,64	0,86
25	η_e, Pa·s	1,40	0,73	0,33	0,09	2,50	1,10	0,87	0,35	7,60	3,75	2,37	1,50
	$b \cdot 10^3$, s^{-1}	1,15	3,37	4,20	32,3	1,66	1,02	2,91	7,31	0,36	0,76	1,50	2,44
	η_f, Pa·s	0,31	0,17	0,14	0,05	1,00	0,62	0,36	0,24	4,95	2,11	2,03	0,68
30	η_e, Pa·s	0,95	0,57	0,25	0,06	1,30	0,76	0,52	0,32	4,05	2,21	1,86	1,00
	$b \cdot 10^3$, s^{-1}	1,38	4,30	5,90	35,0	2,23	1,80	3,14	8,69	0,72	0,83	1,70	2,65
	η_f, Pa·s	0,19	0,13	0,11	0,04	0,68	0,50	0,26	0,19	4,07	1,85	1,45	0,43
35	η_e, Pa·s	0,60	0,39	0,21	0,05	0,90	0,35	0,23	0,22	3,50	1,80	1,59	0,79
	$b \cdot 10^3$, s^{-1}	3,67	5,80	6,37	49,0	2,41	3,56	4,60	9,10	0,88	0,96	1,93	3,20
	η_f, Pa·s	0,17	0,12	0,10	0,04	0,56	0,42	0,22	0,17	2,91	1,46	0,98	0,27
40	η_e, Pa·s	0,40	0,19	0,13	0,03	0,65	0,29	0,15	0,12	2,01	1,39	1,19	0,57
	$b \cdot 10^3$, s^{-1}	5,35	6,60	6,90	73,9	2,67	5,60	5,60	16,8	1,33	1,41	2,27	4,24

where M_0 is the molar weight of the link of a chain. Taking into account the Equations (12) and (13) we have:

$$\rho^* = \rho_0 N^{-4/5} \tag{14}$$

at this:

$$\rho_0 = \frac{M_0}{a^3 N_A} \tag{15}$$

can be called as the density into volume of the monomeric link.

In accordance with the SAWS (prior work by Medvedevskikh [16]) the conformational radius R_m of the polymeric chain into concentrated solutions is greater than into diluted ones and is increased at the polymer concentration ρ increasing. Moreover, not one, but m macromolecules with the same conformational radius are present into the conformational volume R_m^3. This leads to the notion of twisted polymeric chains m–ball for which the conformational volume R_m^3 is general and equally accessible. Since the m–ball is not localized with the concrete polymeric chain, it is the virtual that is by the mathematical notion.

It is followed from the SAWS (prior work by Medvedevskikh [16]):

$$R_m = R_f \cdot m^{1/5} \tag{16}$$

$$m^{1/5} = \left(\frac{\rho}{\rho^*}\right)^{1/2} \quad \text{at} \quad \rho \geq \rho^* \tag{17}$$

thus, it can be written as:

$$R_m = aN\left(\frac{\rho}{\rho_0}\right)^{1/2} \tag{18}$$

The shear modulus μ for the m–ball was determined by the equation (prior work by Medvedevskikh [16]:

$$\mu = 1{,}36\frac{RT}{N_A a^3}\left(\frac{\rho}{\rho_0}\right)^2 \tag{19}$$

and, as it can be seen, does not depend on the length of a chain into the concentrated solutions.

Characteristic time t_m^* of the rotary movement of the m–ball and respectively its shear, in accordance with the prior work by Medvedevskikh [14] is equal to:

$$t_m^* = \frac{4}{7} N^{3,4} \left(\frac{\rho}{\rho_0} \right)^{2,5} L_m \tau_m \tag{20}$$

Let us compare the t_m^* with the characteristic time t_f^* of the rotary movement of Flory ball into diluted solution (prior work by Medvedevskikh [14]):

$$t_f^* = \frac{4}{7} N^{1,4} L_f \tau_f \tag{21}$$

In these equations τ_m and τ_f are characteristic times of the segmental movement of the polymeric chains and L_m and L_f are their form factors into concentrated and diluted solutions respectively. Let us note also, that the Equations (20) and (21) are self-coordinated since at $\rho = \rho^*$ the Equation (20) transforms into the Equation (21). The form factors L_m and L_f are determined by a fact how much strong the conformational volume of the polymeric chain is strained into the ellipsoid of rotation, flattened or elongated one as it was shown by Medvedevskikh [14].

3.3 FRICTIONAL COMPONENT OF THE EFFECTIVE VISCOSITY

In accordance with the data of Table 1 the frictional component of the viscosity η_f strongly depends on a length of the polymeric chains, on their concentration and on the temperature. The all spectrum of η_f dependence on *N*, ρ, and *T* we will be considered as the superposition of the fourth movement forms giving the endowment into the frictional component of the solution viscosity. For the solvent such movement form is the Brownian movement of the molecules that is their translation freedom degree: the solvent viscosity coefficient η_s will be corresponding to this translation freedom degree. The analogue of the Brownian movement of the solvent molecules is the segmental movement of the polymeric chain which is responsible for its translation rotation movements, and also for the shear strain. The viscosity coefficient η_{sm} will be corresponding to this segmental movement of the polymeric chain.

Under the action of a velocity gradient *g* of the hydrodynamic flow the polymeric *m*–ball performs the rotary movement also giving the endowment into the frictional component of the viscosity. In accordance with the superposition principle the segmental movement and the external rotary movement of the polymeric chains will be considered as the independent ones. In this case the external rotary movement of the polymeric chains without taking into account the segmental one is similar to the rotation of *m*–ball with the frozen equilibrium conformation of the all *m* polymeric chains represented into *m*–ball. This corresponds to the inflexible Kuhn's wire model (prior work by Kuhn [17]). The viscosity coefficient η_{pm} will be corresponding to the external rotating movement of the *m*–ball under the action of *g*. The all listed movement forms are enough in order to describe the diluted solutions. However, in a case of the concentrated solutions it is necessary to embed one more movement form, namely, the transference of the twisted between themselves polymeric chain one respectively another in *m*–ball. Exactly such relative movement of the polymeric chains contents into itself the all possible linkages effects. Accordingly to the superposition principle the polymeric chains movement does not depend on the listed movement forms if it does not change the equilibrium conformation of the polymeric chains in *m*–ball. The endowment of such movement form into η_f let us note *via* η_{pz} .

Not all the listed movement forms give the essential endowment into the η_f , however for the generality let us start from the taking into account of the all forms. In such a case the frictional component of a viscosity should be described by the equation:

$$\eta_f = \eta_s(1-\varphi)+(\eta_{sm}+\eta_{pm}+\eta_{pz})\varphi \tag{22}$$

or

$$\eta_f - \eta_s = (\eta_{sm}+\eta_{pm}+\eta_{pz}-\eta_s)\varphi \tag{23}$$

here φ is the volumetric part of the polymer into solution. It is equal to the volumetric part of the monomeric links into *m*–ball, so it can be determined by the ratio:

$$\varphi = \frac{\bar{V}N}{N_A R_m^3} \tag{24}$$

in which $\bar{V}$ is the partial–molar volume of the monomeric link into solution.

Combining the Equations (13)–(18) and (24) we will obtain:

$$\varphi = \frac{\bar{V}\rho}{M_0} \tag{25}$$

The ratio of $M_0/\bar{V}$ should be near to the density ρ_m of the liquid monomer. Assuming of this approximation, $M_0/\bar{V} \approx \rho_m$ we have:

$$\varphi = \frac{\rho}{\rho_m} \tag{26}$$

At the rotation of *m*–ball under the action of *g* the angular rotation rate for any polymeric chain is the same but their links depending on the remoteness from the rotation center will have different linear movement rates. Consequently, in *m*–ball there are local velocity gradients of the hydrodynamic flow. Let g_m represents the averaged upon *m*–ball local velocity gradient of the hydrodynamic flow additional to *g*. Then, the tangential or strain shear σ formed by these gradients g_m and g at the rotation movement of *m*–ball in the medium of a solvent will be equal to:

$$\sigma = \eta_s(g+g_m) \tag{27}$$

However, the measurable strain shear correlates with the well known external gradient *g* that gives another effective viscosity coefficient:

$$\sigma = \eta_{pm} g \tag{28}$$

Comparing the Equation (27) and (28) we will obtain:

$$\eta_{pm} - \eta_s = \eta_s g_m / g \tag{29}$$

Noting

$$\eta_{pm}^0 = \eta_s \cdot g_m / g \tag{30}$$

instead of the Equation (23) we will write:

$$\eta_f - \eta_s = \left(\eta_{sm} + \eta_{pm}^0 + \eta_{pz}\right)\varphi \tag{31}$$

The endowment of the relative movement of twisted polymeric chains in *m*–ball into the frictional component of the viscosity should be in general case depending on a number of the contacts between monomeric links independently to which polymeric chain these links belong. That is why we assume:

$$\eta_{pz} \sim \varphi^2 \tag{32}$$

The efficiency of these contacts or linkages let us estimate comparing the characteristic times of the rotation (shear) of *m*–ball into concentrated solution t_m^* and polymeric ball into diluted solution t_f^* determined by the Equations (20) and (21).

We note that in accordance with the determination done by Medvedevskikh [14] t_m^* is characteristic time not only for *m*–ball rotation, but also for each polymeric chain in it. Consequently, t_m^* is the characteristic time of the rotation of polymeric chain twisted with others chains whereas t_f^* is the characteristic time of free polymeric chain rotation. It permits to assume the ratio t_m^*/t_f^* as a measure of the polymeric chains contacts or linkages efficiency and to write the following in accordance with the Equations (20) and (21):

$$\eta_{pz} \sim \frac{t_m^*}{t_f^*} = N^2 \left(\frac{\rho}{\rho_0}\right)^{2,5} \left(\frac{L_m \tau_m}{L_f \tau_f}\right) \tag{33}$$

Taking into account the Equation (26) and combining the Equations (32) and (33) into one equation we will obtain:

$$\eta_{pz} = \eta_{pz}^0 N^2 \left(\frac{\rho}{\rho_0}\right)^{2,5} \left(\frac{\rho}{\rho_m}\right)^2 \tag{34}$$

Here the coefficient of proportionality η_{pz}^0 includes the ratio $L_m \tau_m / L_f \tau_f$, which should considerably weaker depends on ρ and *N* that the value η_{pz}.

Substituting the Equation (34) into (31) with taking into account the Equation (26) we have:

$$\eta_f - \eta_s = \left[\eta_{sm} + \eta_{pm}^0 + \eta_{pz}^0 N^2 \left(\frac{\rho}{\rho_0}\right)^{2,5} \left(\frac{\rho}{\rho_m}\right)^2\right] \frac{\rho}{\rho_m} \tag{35}$$

Let us estimate the endowment of the separate terms in Equation (35) into η_f. In accordance with Table 1 under conditions of experiments the frictional component of the viscosity is changed from the minimal value ≈ 4×10^{-2} Pa×s to the maximal one ≈ 6.5 Pa×s. Accordingly to the reference data the viscosity coefficient η_s of the toluene has the order 5×10^{-4} Pa×s. The value of the viscosity coefficient η_{sm} representing the segmental movement of the polymeric chains estimated by us upon η_f of the diluted solution of polystyrene in toluene consists of the value by 5×10^{-3} Pa×s order. Thus, it can be assumed η_{sm}, $\eta_s << \eta_f$ and it can be neglected the respective terms in Equation (35). With taking into account of this fact, the Equation (35) can be rewritten in a form convenient for the graphical test:

$$\eta_f \frac{\rho_m}{\rho} = \eta_{pm}^0 + \eta_{pz}^0 N^2 \left(\frac{\rho}{\rho_0}\right)^{2,5} \left(\frac{\rho}{\rho_m}\right)^2 \tag{36}$$

On Figure 2, it is presented the interpretation of the experimental values η_f into coordinates of the Equation (36).

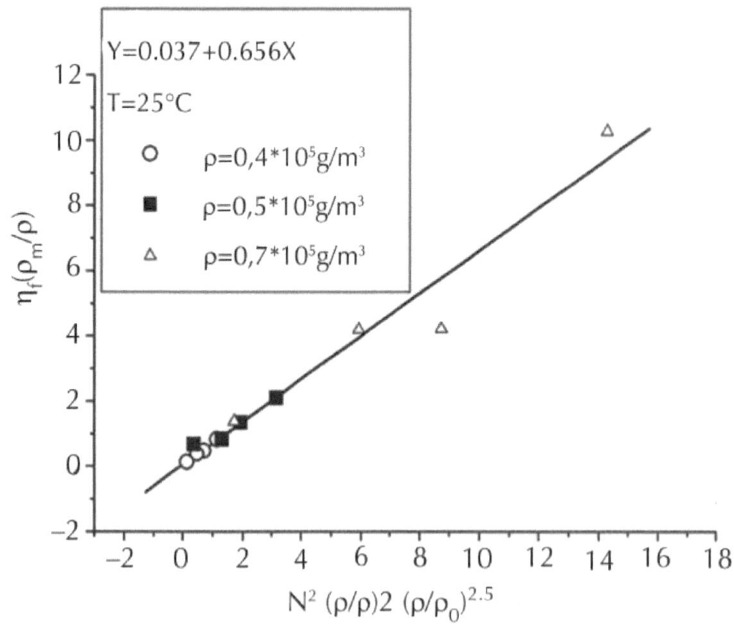

FIGURE 2 *(Continued)*

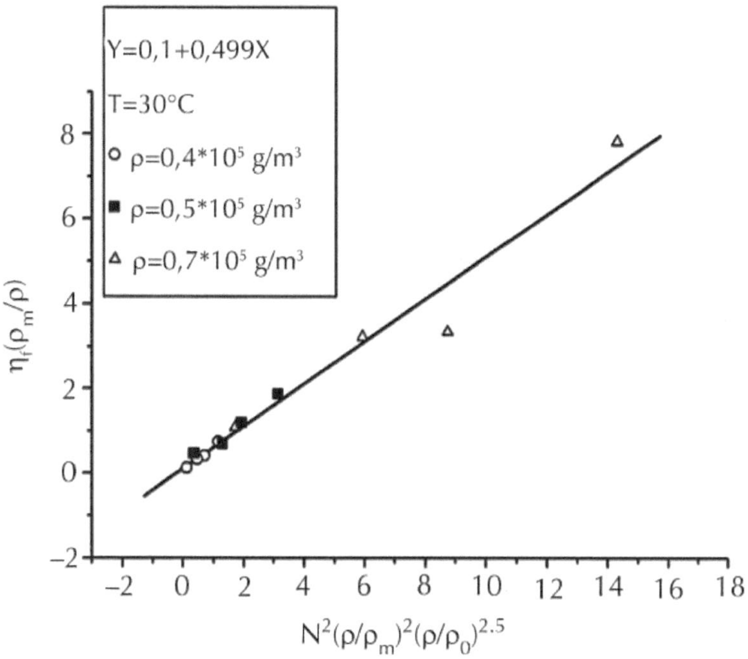

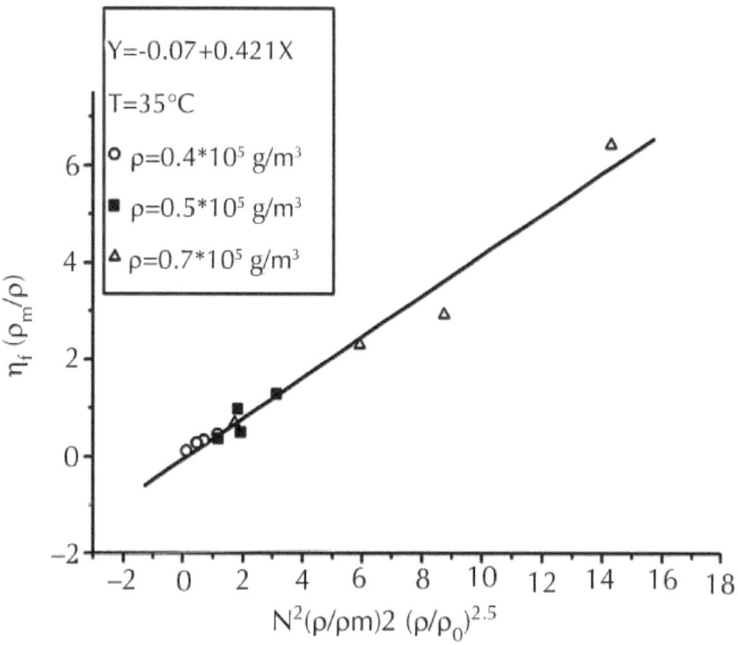

FIGURE 2 *(Continued)*

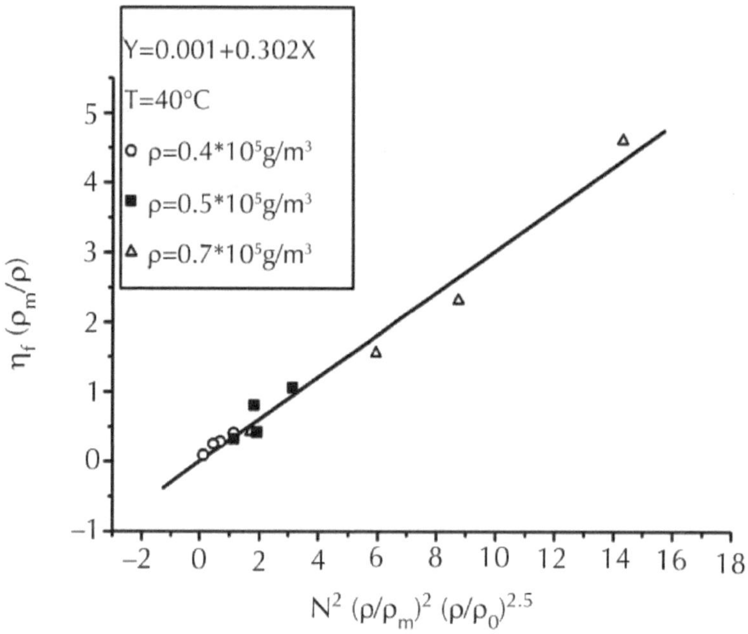

FIGURE 2 Interpretation of the experimental values of friction viscosity η_f in coordinates of Equation (32).

At that, it were assumed the following values: M = 104,15 g/mol, a = 1,86×10^{-10} m under determination of ρ_0 accordingly to Equation (15) and $\rho_m = 0,906 \cdot 10^6$ g/m^3 for liquid styrene. As we can see, the linear dependence is observed corresponding to Equation (36) at each temperature, based on the tangent of these straight lines inclination (see the regression equations on Figure 2) it were found the numerical values of η^0_{pz}, the temperature dependence of which is shown on Figure 3 into the Arrhenius' coordinates.

It is follows from these data, that the activation energy E_{pz} regarding to the movement of twisted polymeric chains in toluene is equal to 39.9 kJ/mol.

It can be seen from the Figure 2 and from the represented regression equations on them, that the values η^0_{pm} are so little (probably, $\eta^0_{pm} << 0,1$ $Pa{\times}s$) that they are located within the limits of their estimation error. This, in particular, did not permit us to found the numerical values of the ratio g_m / g.

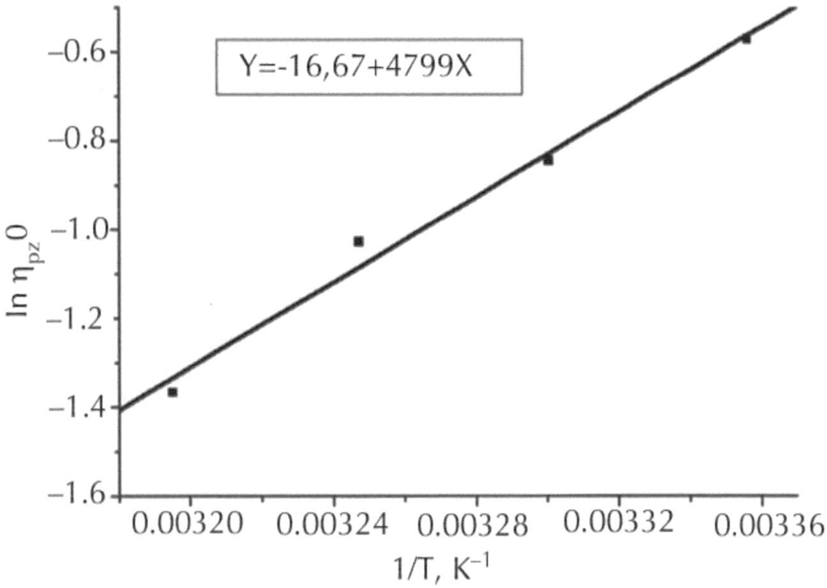

FIGURE 3 Temperature dependence of the coefficient of viscosity η_{pz}^{0} in the Arrhenius equation plots.

So, the analysis of experimental data, which has been done by us, showed that the main endowment into the frictional component of the effective viscosity of the concentrated solutions "polystyrene in toluene" has the separate movement of the twisted between themselves into *m*–ball polymeric chains. Exactly this determines a strong dependence of the η_f on concentration of polymer into solution $\left(\eta_f \sim \rho^{5,5}\right)$ and on the length of a chain $\left(\eta_f \sim N^2\right)$.

3.4 ELASTIC COMPONENT OF THE EFFECTIVE VISCOSITY

It is follows from the data of Table 1, that the elastic component of viscosity η_e is a strong increasing function on polymer concentration ρ, on a length of a chain *N*, and a diminishing function on a temperature *T*.

The elastic properties of the conformational state of the *m*–ball of polymeric chains are appeared in a form of the resistance to the conformational volume deformation under the action of the external forces. In particular, the resistance to the shear is determined by the shear modulus μ, which for the *m*–ball was determined by the Equation (19).

As it was shown by Medvedevskikh [14], the elastic component of the viscosity is equal to:

$$\eta_e = \mu t_m^* L_m \tag{37}$$

The factor of form L_m depends on the deformation degree of the conformational volume of a ball. In particular, at the deformation of a m–ball with the conformational radius R_m into the rotation ellipsoid (flattened out or elongated, for example, along the z axis), the factor of a form is determined by the equation proposed by Medvedevskikh [14, 16].

$$L_m = \frac{3}{\lambda_v^2\left(\frac{1}{\lambda_x} + \frac{4}{(\lambda_x + \lambda_z)}\right)} \tag{38}$$

Here $\lambda_x = \lambda_y$ and λ_z are multiplicities of the linear deformation of m–ball along the corresponding axes $\lambda_i = R_i / R_m$, where R_i is a semi-axis of the ellipsoid, λ_v is the multiplicity of the volumetric deformation: $\lambda_v = \lambda_x \lambda_y \lambda_z = \lambda_x^2 \lambda_z$.

Combining the Equations (19) and (20) into (37) and assuming $\frac{4}{7} \cdot 1{,}36 \approx 1$ we will obtain:

$$\eta_e = \frac{RT}{M_0} N^{3,4} \rho \left(\frac{\rho}{\rho_0}\right)^{3,5} L_m \tau_m \tag{39}$$

Comparing the Equation (20) and (39) we can see, that the known from the reference data ratio $\eta_e \sim t_m^* \sim N^{3,4}$ is performed but only for the elastic component of a viscosity.

It is follows from the Equation (39), that the parameters L_m and τ_m are inseparable, so, based on the experimental values of η_e (see Table 1) it can be found the numerical values only for the composition $L_m \cdot \tau_m$. The results of $(L_m \tau_m)\eta$ calculations are represented in Table 2.

In spite of these numerical estimations scattering it is overlooked their clear dependence on T, but not on ρ and N.

3.5 PARAMETER B

In accordance with the determination Equation (9), the b parameter is a measure of the velocity gradient of hydrodynamic flow created by the working cylinder rotation, influence on characteristic time t_v^* of g action on the shear strain of the m–ball and its rotation movement. Own characteristic time t_m^* of m–ball shear and rotation accordingly to Equation (20) depends only on ρ, N, and T *via* τ_m.

It is follows from the experimental data (see Table 1) that the b parameter is a function on the all three variables ρ, N, and T, but, at that, is increased at T increasing and is decreased at ρ and N increasing. In order to describe these dependences let us determine the angular rate ω_m^0 (s^{-1}) of the strained m–ball rotation with the effective radius $R_m L_m$ of the working cylinder by diameter d contracting with the surface:

$$\omega_m^0 = \frac{\pi d \omega}{R_m L_m} \tag{40}$$

TABLE 2 Calculated values Lτ, τ/L, τ and L based on the experimental magnitudes η_e and b.

$\rho \cdot 10^{-5}$, g/m^3		4,0				5,0				7,0				
T,°C	$M \cdot 10^4$, g/mole	5,1	4,1	3,3	2,2	5,1	4,1	3,3	2,2	5,1	4,1	3,3	2,2	$\tilde{\tau} 10^{10}$, s $\tilde{L}$
25	$(L\tau)_{\eta e} \cdot 10^{10}$, s	2,63	3,14	2,72	2,99	1,71	1,72	2,61	4,25	1,15	1,29	1,57	4,00	
	$(\tau/L)_b \cdot 10^{10}$, s	3,25	1,81	2,54	0,89	1,17	3,43	1,91	2,06	1,98	1,86	1,38	2,29	
	$\tau \cdot 10^{10}$, s	2,92	2,38	2,63	1,63	1,41	2,43	2,23	2,96	1,51	1,61	1,47	3,03	2,19
	L	0,90	1,32	1,03	1,83	1,21	0,71	1,17	1,44	0,76	0,86	1,07	1,32	1,13
30	$(L\tau)_{\eta e} \cdot 10^{10}$, s	1,75	2,41	2,03	1,96	0,88	1,17	1,54	3,83	0,60	0,75	1,21	2,63	
	$(\tau/L)_b \cdot 10^{10}$, s	2,10	1,56	1,81	0,82	0,87	1,94	1,39	1,73	1,00	1,62	1,22	2,11	
	$\tau \cdot 10^{10}$, s	2,17	1,94	1,92	1,27	0,88	1,51	1,46	2,57	0,78	0,98	1,21	2,56	1,59
	L	0,81	1,24	1,00	1,55	1,00	0,78	1,05	1,49	0,78	0,60	1,00	1,12	1,04
35	$(L\tau)_{\eta e} \cdot 10^{10}$, s	1,09	1,62	1,67	1,61	0,60	0,53	0,67	2,58	0,51	0,60	1,02	2,04	
	$(\tau/L)_b \cdot 10^{10}$, s	1,01	1,16	1,67	0,59	0,79	0,98	1,21	1,65	0,81	1,35	1,09	1,75	
	$\tau \cdot 10^{10}$, s	1,05	1,37	1,67	0,97	0,70	0,72	0,90	2,06	0,64	0,90	1,05	1,89	1,16
	L	1,04	1,18	1,00	1,65	0,87	0,73	0,74	1,25	0,79	0,67	0,97	1,08	1,00

TABLE 2 *(Continued)*

$\rho \cdot 10^{-5}$, g/m^3		4,0				5,0				7,0				$\tilde{\tau} 10^{10}$, s $\tilde{L}$
$T, ^0C$	$M \cdot 10^4$, g/mole	5,1	4,1	3,3	2,2	5,1	4,1	3,3	2,2	5,1	4,1	3,3	2,2	
40	$(L\tau)_{\eta e} \cdot 10^{10}$, s	0,72	0,78	1,03	0,96	0,43	0,44	0,43	1,40	0,29	0,46	0,75	1,46	
	$(\tau/L)_b \cdot 10^{10}$, s	0,70	1,01	1,54	0,39	0,73	0,62	1,00	0,90	0,54	0,92	0,91	1,31	
	$\tau \cdot 10^{10}$, s	0,71	0,89	1,26	0,61	0,56	0,52	0,66	1,12	0,40	0,65	0,83	1,38	0,80
	L	1,01	0,88	0,82	1,57	0,77	0,84	0,66	1,25	0,73	0,71	0,91	1,06	0,93

here π is appeared due to the difference in the dimensionalities of ω_m^0 and ω.

Let us determine the t_v^0 as the reverse one ω_m^0:

$$t_v^0 = \frac{R_m L_m}{\pi d\omega} \quad (41)$$

According to Equation (41) t_v^0 is a time during which the m–ball with the effective radius $R_m L_m$ under the action of working cylinder by diameter d rotation will be rotated on the angle equal to the one radian. Let us note, that the t_m^* was determined by Medvedevskikh [14] also in calculation of the m–ball turning on the same single angle.

Since in experiments the working cylinder had two rotating surfaces with the diameters d_1 and d_2, the value ω_m^0 was averaged out in accordance with the condition $d = (d_1 + d_2)/2$, so, respectively, the value t_v^0 was averaged out too:

$$t_v^0 = \frac{2R_m L_m}{\pi(d_1 + d_2)\omega} \quad (42)$$

So, t_v^0 is in inverse proportion to ω, therefore through the constant device it is in inverse proportion to g: $t_v^0 \sim g^{-1}$. However, as it was noted, in m–ball due to the difference in linear rates of the polymeric chains links it is appeared the hydrodynamic interaction which leads to the appearance of the additional to g local averaged upon m–ball velocity gradient of the hydrodynamic flow g_m. This local gradient g_m acts not on the conformational volume of the m–ball but on the monomeric framework of the polymeric chains (the inflexible Kuhn's wire model (prior work by Kuhn [17]). So, the endowment of g_m into characteristic time t_v^* depends on the volumetric part φ of the links into the conformational volume of m–ball, that is $t_v^* \sim (g + g_m\varphi)^{-1}$.

Therefore, it can be written the following:

$$\frac{t_v^*}{t_v^0} = \frac{g}{g + g_m\varphi} \quad (43)$$

that with taking into account of Equation (42) leads to the equation:

$$t_v^* = \frac{2R_m L_m}{\pi(d_1 + d_2)\omega} \bigg/ \left(1 + \frac{g_m}{g}\frac{\rho}{\rho_m}\right) \quad (44)$$

Combining the Equations (20) and (44) into Equation (9) we will obtain:

$$b = \frac{7a}{2\pi(d_1 + d_2)} \cdot \frac{L_m}{\tau_m} / N^{2,4} \left(\frac{\rho}{\rho_0}\right)^2 \left(1 + \frac{g_m}{g}\frac{\rho}{\rho_m}\right) \quad (45)$$

As we can see, here the parameters L_m and τ_m are also inseparable and cannot be found independently one from another. That is why based on the experimental data presented in Table 1 it can be found only the numerical values of the ratio $(\tau_m / L_m)_b$. After the substitution of values $a = 1.86 \times 10^{-10}$ m, $d_1 = 3.4 \times 10^{-2}$ m, and $d_2 = 3.3 \times 10^{-2}$ m we have:

$$\left(\frac{\tau_m}{L_m}\right)_b = 2,84 \cdot 10^{-9} / N^{2,4} \left(\frac{\rho}{\rho_0}\right)^2 \left(1 + \frac{g_m}{g}\frac{\rho}{\rho_m}\right) b \tag{46}$$

As it was marked, we could not estimate the numerical value of g_m / g due to the smallness of the value η^0_{pm} lying in the error limits of its measuring. That is why, we will be consider the ratio g_m / g as the fitting parameter starting from the consideration that the concentrated solution for polymeric chains is more ideal than the diluted one and moreover, the *m*–ball is less strained than the single polymeric ball. That is why, g_m / g was selected in such a manner that the factor of form L_m was near to the 1. This lead to the value g_m / g =25.

The calculations results of $(\tau_m / L_m)_b$ accordingly to Equation (46) with the use of experimental values from Table 1 and also the values g_m / g =25 are represented in Table 2. They mean that the $(\tau_m / L_m)_b$ is a visible function on a temperature but not on a ρ and *N*.

3.6 CHARACTERISTIC TIME OF THE SEGMENTAL MOVEMENT

On a basis of the independent estimations of $(\tau_m / L_m)_\eta$ and $(\tau_m / L_m)_b$ it was found the values of τ_m and L_m, which also presented in Table 2. An analysis of these data shows that with taking into of their estimation error it is discovered the clear dependence of τ_m and *L* on *T*, but not on ρ and *N*. Especially clear temperature dependence is visualized for the values $\tilde{\tau}_m$, obtained *via* the averaging of τ_m at giving temperature for the all values of ρ and *N* (Table 2).

The presentation of values $\tilde{\tau}_m$ into Arrhenius' coordinate's equation (Figure 4) permitted to obtain the equation:

$$\ln \tilde{\tau}_m = -42,23 + 5950 / T \tag{47}$$

From which it is follows that the activation energy $E_{\tau m}$ of the segmental movement for polymeric chains into *m*–ball is equal to 49.44 kJ/mol.

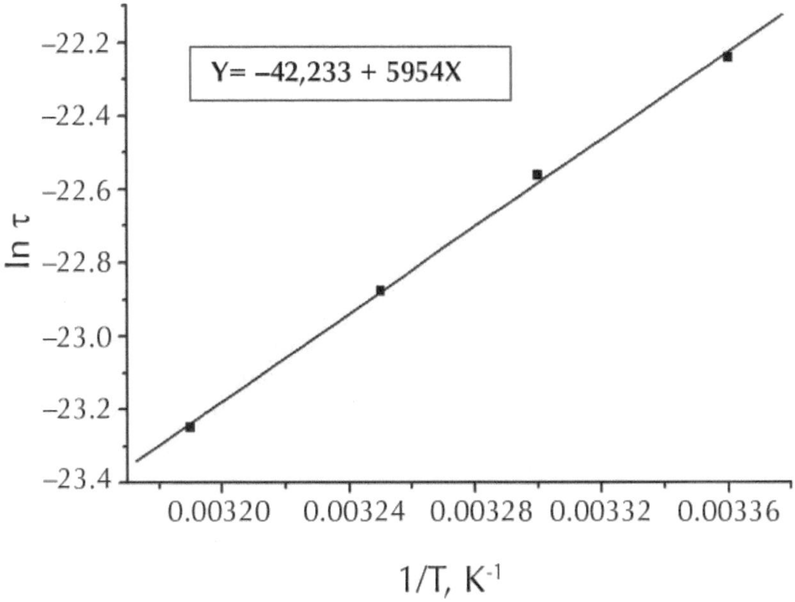

FIGURE 4 Temperature dependence of the average values of characteristic time $\tilde{\tau}$ of the segmental movement of polymeric chain in the Arrhenius equation plots.

Early it was obtained some other equation for the temperature dependence of the characteristic time τ_f of segmental movement for diluted solution of polystyrene in toluene:

$$\ln \tilde{\tau}_f = -44,07 + 6660 / T \tag{48}$$

that gives the value of activation energy $E_{\tau f}$ equal to 55,34 kJ/mol.

Via presented temperature interval $\tilde{\tau}_m < \tilde{\tau}_f$ is approximately in 1,5 times, the difference into values of $E_{\tau m}$ and $E_{\tau f}$ for concentrated and diluted solutions is not too large also. So, the difference in the Equations (47) and (48) can be quite acceptably taken to the error of the experimental estimation of $\tilde{\tau}_m$ and $\tilde{\tau}_f$. From the other hand, the ratio $\tilde{\tau}_m < \tilde{\tau}_f$ and $E_{\tau m} < E_{\tau f}$ can be explained by fact that due to the thermodynamic properties the concentrated solution is more ideal for polymeric chains than the diluted one.

Let us compare the values $E_{\tau m}$ and $E_{\tau f}$ with the evaporation heats ΔH_{evap} of styrene (–43.94 kJ/mol) and toluene (–37.99 kJ/mol). As we can see, independently on fact which values of ΔH_{evap} can be taken, it is observed the general picture $E_{\tau m}, E_{\tau f} > \Delta H_{evap}$ for styrene or toluene. At the same time, it is known from prior works Gleston et al. [6]

that for low–molecular liquids the viscosity of which is determined by the Brownian or translational form of the molecules movement, the activation energy of the viscous flow is in 3–4 times less than the evaporation heat. This means that the segmental movement which is a base of the reptation mechanism of the polymeric chains transfer is determined by their strain vibrating freedom degrees.

3.7 CONCLUSION

Investigation of the effective viscosity gradient dependence for concentrated solution of polystyrene in toluene permitted to mark out its frictional η_f and elastic η_e components and to study their dependence on the concentration ρ of polymer into solution, on the length of a polymeric chain *N* and on a temperature *T*. It was determined that the main endowment into the frictional component of the viscosity has the relative movement of the twisted between themselves into *m*–ball polymeric chains. The efficiency of the all possible at this linkages was determined by the ratio of the characteristic times of rotation movement of polymeric chains into concentrated t_m^* and diluted t_f^* solutions. This lead to the dependence $\eta_f \sim N^2\rho^{5,5}$ which is conforming with the experimental data.

The elastic component of the viscosity η_e is determined by the elastic properties of the conformational volume of *m*–ball of polymeric chains at its shear strain. Due to the kinetic reasons the endowment of η_e in η depends on the velocity gradient of the hydrodynamic flow. The measure of this dependence is the *b* parameter which is also estimated based on the experimental data.

It was experimentally confirmed the determined early theoretical dependence of $\eta_e \sim N^{3,4}\rho^{4,5}$ that leads to the known ratio $\eta_e \sim t_m^* \sim N^{3,4}$ which is true, however, only for the elastic component of a viscosity. On a basis of the experimental values η_e and *b* it were obtained the numerical values of the characteristic time τ_m of the segmental movement and the factor of form L_m of *m*–ball. These data show that the τ_m does not depend on ρ and *N*, but only on the temperature. On a basis of the averaged values of $\tilde{\tau}_m$ at all ρ and *N* at given *T* it was found the activation energy of the segmental movement. This value is equal to $E_{\tau m}$ = 49.44 kJ/mol and exceeds the evaporation heat both of the solvent and the monomer. It can be concluded that the segmental movement is determined by the strain vibrating freedom degrees of the polymeric chains.

KEYWORDS

- **Activation energy**
- **Parameter b**
- **Effective viscosity**
- **M–ball**
- **Segmental movement**

REFERENCES

1. Ferry, L. D. *Viscoelastic Properties of Polymers*, 2nd Ed.. John Willey & Sons, New York (1970).
2. De Jennes, P. G. *Scaling Concepts in Polymer Physics*. Ithaca, Cornell Univ. Press, p. 300 (1979).
3. Tsvietkov, V. N., Eskin, V. E., and Frenkiel, S. Ya. *The Structure of Macromolecules Into Solutions*. (in Russian), Nauka, Moscow, p. 700 (1964).
4. Malkin, A. Ya. and Isayev, A. I. *Rheology Conception, Methods, Applications*. (in Russian), Chem. Tech. Publishing, Toronto, Canada p. 465 (2005).
5. Grassley, W. The entanglement concept in polymer rheology. *Adv. Polym. Sci.*, **16**, 1–8 (1974).
6. Glesston, S., Leidlier, K., and Eiring, G. *Theory of Absolute Reaction Rates*. (in Russian), M. "Inostr. Lit.", p. 583 (1948).
7. Peterlin, A. and Copic, M. Gradient dependence of the intrinsic viscosity of linear macromolecules. *J. Appl. Phys.*, **27**, 434–439 (1956).
8. Hoffman, M. and Rother, R. Strukturviskositat und molekulare struktur von Fadenmolekulen. *Macromol. Chem.*, **80**, 95–111 (1964).
9. Leonov, A. I. and Vinogradov, G. V. *Reports of the Academy of Sciences of USSR*. (in Russian), **155**(2), 406–409 (1964).
10. Williams, M. C. Concentrated Polymer Solutions Part II. Dependence of Viscosity and Relaxation Time on Concentration and Molecular Weight. *A. I. Ch. E. Journal*, **13**, 534–539 (1967).
11. Bueche, F. Viscosity of polymers in concentrated solution. *Journ. Chem. Phys.*, **25**, 599–605 (1956).
12. Edwards, S. F. and Grant, J. W. The effect of entanglements of diffusion in a polymer melt. *Journ. Phys.*, **46**, 1169–1186 (1973).
13. De Jennes, P. G. Reptation of a Polymer Chain in the Presence of Fixed Obstacles. *Journ. Chem. Phys.*, **55**, 572–580 (1971).
14. Medvedevskikh, Yu. G. *Conformation of Macromolecules Thermodynamic and Kinetic Demonstrations*. Nova Sci. Publishing, New York, p. 249 (2007).
15. Medvedevskikh, Yu. G. Statistics of linear polymer chains in the self-avoiding walksnmodel. *Condensed Matter Physics*, **2**(26), 209–218 (2001).
16. Medvedevskikh, Yu. G. Conformation and Deformation of Linear Macromolecules in Dilute Ideal Solution in the Self–Avoiding Random Walks Statistics. *J. Appl. Polymer Sci.*, **109**, 2472–2481 (2008).
17. Kuhn, H. and Kuhn, W. Effects of Hampered Draining of Solvent on the Translatory and Rotatory Motion of Statistically Coiled Long-Chain Molecules in Solution Part II. Rotatory Motion, Viscosity, and Flow Birefringence. *J. Polymer Sci.*, **9**, 1–33 (1952).

4 Update on Reinforcement Mechanisms of Polymer Nanocomposites

G. V. Kozlov, Yu. G. Yanovskii, and G. E. Zaikov

CONTENTS

4.1 INTRODUCTION

The modern methods of experimental and theoretical analysis of polymer materials structure and properties allow not only to confirm earlier propounded hypotheses, but to obtain principally new results. Let us consider some important problems of particulate-filled polymer nanocomposites, the solution of which allows advancing substantially in these materials properties understanding and prediction. Polymer nanocomposites multicomponentness (multiphaseness) requires their structural components quantitative characteristics determination. In this aspect interfacial regions play a particular role, since it has been shown earlier, that they are the same reinforcing element in elastomeric nanocomposites as nanofiller actually [1]. Therefore the knowledge of interfacial layer dimensional characteristics is necessary for quantitative determination of one of the most important parameters of polymer composites in general – their reinforcement degree [2, 3].

The aggregation of the initial nanofiller powder particles in more or less large particles aggregates always occur in the course of technological process of making particulate-filled polymer composites in general [4] and elastomeric nanocomposites in particular [5]. The aggregation process tells on composites (nanocomposites) macroscopic properties [2-4]. For nanocomposites nanofiller aggregation process gains special significance, since its intensity can be one, that nanofiller particles aggregates size exceeds 100 nm – the value, which is assumed (though conditionally enough [6]) as an upper dimensional limit for nanoparticle. In other words, the aggregation process can result to the situation when primordially supposed nanocomposite ceases to be one. Therefore at present several methods exist, which allow to suppress nanoparticles aggregation process [5, 7]. This also assumes the necessity of the nanoparticles aggregation process quantitative analysis.

It is well-known [1, 2], that in particulate-filled elastomeric nanocomposites (rubbers) nanofiller particles form linear spatial structures ("chains"). At the same time in polymer composites, filled with disperse microparticles (microcomposites) particles (aggregates of particles) of filler form a fractal network, which defines polymer matrix structure (analog of fractal lattice in computer simulation) [4]. This results to different mechanisms of polymer matrix structure formation in micro and nanocomposites. If in the first filler particles (aggregates of particles) fractal network availability results to "disturbance" of polymer matrix structure, that is expressed in the increase of its fractal dimension d_f [4], then in case of polymer nanocomposites at nanofiller contents change the value d_f is not changed and equal to matrix polymer structure fractal dimension [3]. As it has been expected, the change of the composites of the indicated classes structure formation mechanism change defines their properties, in particular, reinforcement degree [11, 12]. Therefore nanofiller structure fractality strict proof and its dimension determination are necessary.

As it is known [13, 14], the scale effects in general are often found at different materials mechanical properties study. The dependence of failure stress on grain size for metals (Holl-Petsch formula) [15] or of effective filling degree on filler particles size in case of polymer composites [16] are examples of such effect. The strong dependence of elasticity modulus on nanofiller particles diameter is observed for particulate-filled elastomeric nanocomposites [5]. Therefore it is necessary to elucidate the physical grounds of nano and micromechanical behavior scale effect for polymer nanocomposites.

At present a disperse material wide list is known, which is able to strengthen elastomeric polymer materials [5]. These materials are very diverse on their surface chemical constitution, but particles small size is a common feature for them. On the basis of this observation the hypothesis was offered, that any solid material would strengthen the rubber at the condition, that it was in a much dispersed state and it could be dispersed in polymer matrix. Edwards [5] points out, that filler particles small size is necessary and, probably, the main requirement for reinforcement effect realization in rubbers. Using modern terminology, one can say, that for rubbers reinforcement the nanofiller particles, for which their aggregation process is suppressed as far as possible would be the most effective ones [3, 12]. Therefore, the theoretical analysis of a nanofiller particles size influence on polymer nanocomposites reinforcement is necessary.

Proceeding from chapter, the present work purpose is the solution of the considered paramount problems with the help of modern experimental and theoretical techniques on the example of particulate-filled butadiene-styrene rubber.

4.2 EXPERIMENTAL

The made industrially butadiene-styrene rubber of mark SKS-30, which contains 7.0–12.3% cis- and 71.8–72.0% trans bonds, with density of 920–930 kg/m^3 was used as matrix polymer. This rubber is fully amorphous one.

The fullerene containing mineral shungite of Zazhoginsk's deposit consists of ~30% globular amorphous metastable carbon and ~70% high-disperse silicate particles. Besides, industrially made technical carbon of mark № 220 was used as nanofiller. The technical carbon, nano and microshugite particles average size makes up 20, 40, and 200 nm, respectively. The indicated filler content is equal to 37 mass%. Nano and microdimensional disperse shungite particles were prepared from industrially output material by the original technology processing. The size and polydispersity analysis of the received in milling process shungite particles was monitored with the aid of analytical disk centrifuge (CPS Instruments, Inc., USA), allowing to determine with high precision size and distribution by the sizes within the range from 2 nm up to 50 mcm.

Nanostructure was studied on atomic forced microscopes Nano-department of science and technology (DST) (Pacific Nanotechnology, USA) and Easy Scan dynamic force microscope (DFM) (Nanosurf, Switzerland) by semi-contact method in the force modulation regime. Atomic force microscopy (AFM) results were processed with the help of specialized software package scanning probe image processor (SPIP, Denmark). The SPIP is a powerful programs package for processing of images, obtained on scanning probe microscopy (SPM), AFM, scanning tunneling microscopy (STM), scanning electron microscopes (SEM), transmission electron microscopes (TEM), interferometers, confocal microscopes, profilometers, optical microscopes, and so on. The given package possesses the whole functions number, which are necessary at images precise analysis, in a number of which the following ones are included:

(1) The possibility of three-dimensional reflecting objects obtaining, distortions automatized leveling, including Z-error mistakes removal for examination of separate elements and so on.
(2) Quantitative analysis of particles or grains, more than 40 parameters can be calculated for each found particle or pore: area, perimeter, mean diameter, the ratio of linear sizes of grain width to its height distance between grains, coordinates of grain center of mass (a, a) can be presented in a diagram form or in a histogram form.

The tests on elastomeric nanocomposites nanomechanical properties were carried out by a nanoindentation method [17] on apparatus Nano Test 600 (Micro Materials, Great Britain) in loads wide range from 0.01 mN up to 2.0 mN. Sample indentation was conducted in 10 points with interval of 30 mcm. The load was increased with constant rate up to the greatest given load reaching (for the rate 0.05 mN/s-1 mN). The indentation rate was changed in conformity with the greatest load value counting, that loading cycle should take 20 s. The unloading was conducted with the same rate as loading. In the given experiment the "Berkovich indentor" was used with the angle at the top of 65.3° and rounding radius of 200 nm. The indentations were carried out in the checked load regime with preload of 0.001 mN.

For elasticity modulus calculation the obtained in the experiment by nanoindentation course dependences of load on indentation depth (strain) in ten points for each sample at loads of 0.01, 0.02, 0.03, 0.05, 0.10, 0.50, 1.0, and 2.0 mN were processed according to Oliver-Pharr method [18].

4.3 DISCUSSION AND RESULTS

In Figure 1 the obtained according to the original methodics results of elasticity moduli calculation for nanocomposite butadiene-styrene rubber/nanoshungite components (matrix, nanofiller particle and interfacial layers), received in interpolation process of nanoindentation data are presented. The processed in SPIP polymer nanocomposite image with shungite nanoparticles allows experimental determination of interfacial layer thickness l_{if}, which is presented in Figure 1 as steps on elastomeric matrix-nanofiller boundary. The measurements of 34 such steps (interfacial layers) width on the processed in SPIP images of interfacial layer various section gave the mean experimental value l_{if} = 8.7 nm. Besides, nanoindentation results (Figure 1, figures on the right) showed, that interfacial layers elasticity modulus was only by 23–45% lower than nanofiller elasticity modulus, but it was higher than the corresponding parameter of polymer matrix in 6.0–8.5 times. These experimental data confirm, that for the studied nanocomposite interfacial layer is a reinforcing element to the same extent as nanofiller actually [1, 3, 12].

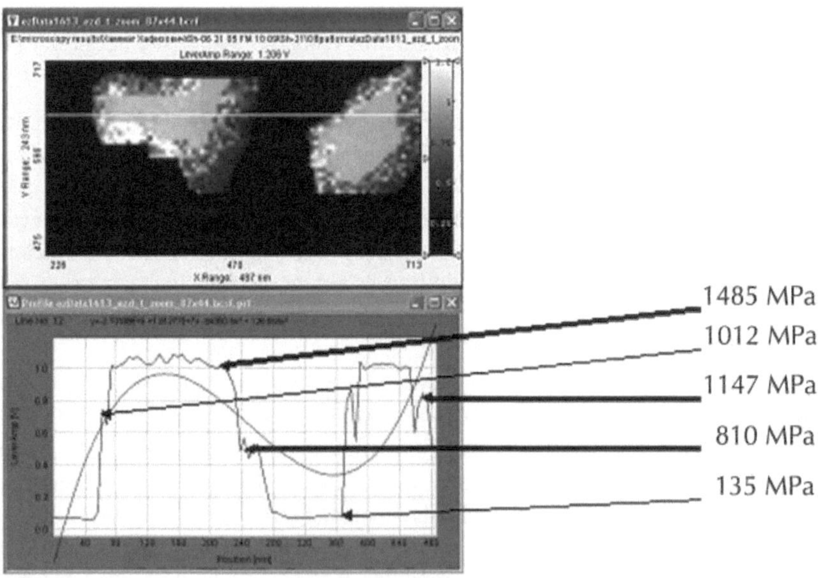

FIGURE 1 The processed in SPIP image of nanocomposite butadiene-styrene rubber/nanoshungite, obtained by force modulation method and mechanical characteristics of structural components according to the data of nanoindentation (strain 150 nm).

Let us fulfill further the value l_{if} theoretical estimation according to the two methods and compare these results with the ones obtained experimentally. The first method simulates interfacial layer in polymer composites as a result of interaction of two fractals–polymer matrix and nanofiller surface [19, 20]. In this case there is a sole linear scale l, which defines these fractals interpenetration distance [21]. Since nanofiller elasticity modulus is essentially higher, than the corresponding parameter for rubber in the considered case–in 11 times (see Figure 1), then the indicated interaction reduces to nanofiller indentation in polymer matrix and then $l = l_{if}$. In this case it can be written [21]:

$$l_{if} \approx a\left(\frac{R_p}{a}\right)^{2(d-d_{surf})/d} \quad (1)$$

where a is a lower linear scale of fractal behavior, which is accepted for polymers as equal to statistical segment length l_{st} [22], R_p is a nanofiller particle (more precisely, particles aggregates) radius, for which nanoshungite is equal to ~84 nm [23], d is dimension of Euclidean space, in which fractal is considered (it is obvious, that in case $d = 3$), d_{surf} is fractal dimension of nanofiller particles aggregate surface. The value l_{st} is determined as follows [24]:

$$l_{st} = l_0 C_\infty , \quad (2)$$

where l_0 is the main chain skeletal bond length, which is equal to 0.154 nm for both blocks of butadiene-styrene rubber [25], C is characteristic ratio, which is a polymer chain statistical flexibility indicator [26], and is determined with the help of the equation [22]:

$$T_g = 129\left(\frac{S}{C_\infty}\right)^{1/2} , \quad (3)$$

where T_g is glass transition temperature, equal to 217K for butadiene-styrene rubber [3], S is macromolecule cross-sectional area, determined for the mentioned rubber according to the additivity rule from the following considerations. As it is known [27], the macromolecule diameter quadrate values are equal: for polybutadiene – 20.7 Å^2 and for polystyrene – 69.8 Å^2. Having calculated cross-sectional area of macromolecule, simulated as a cylinder, for the indicated polymers according to the known geometrical formulas, let us obtain 16.2 and 54.8 Å^2, respectively. Further, accepting as S the average value of the adduced above areas, let us obtain for butadiene-styrene rubber $S = 35.5$ Å^2. Then according to the Equation (3) at the indicated values T_g and S let us obtain $C_\infty = 12.5$ and according to the Equation (2) – $l_{st} = 1.932$ nm.

The fractal dimension of nanofiller surface d_{surf} was determined with the help of the equation [3]:

$$S_u = 410 R_p^{d_{surf}-d} , \quad (4)$$

where S_u is nanoshungite particles specific surface, calculated as follows [28]:

$$S_u = \frac{3}{\rho_n R_p}, \tag{5}$$

where ρ_n is the nanofiller particles aggregate density, determined according to the formula [3]:

$$\rho_n = 0.188\left(R_p\right)^{1/3} \tag{6}$$

The calculation according to the Equations (4)–(6) gives d_{surf} = 2.44. Further, using the calculated by the indicated mode parameters, let us obtain from the Equation (1) the theoretical value of interfacial layer thickness l_{if}^T = 7.8 nm. This value is close enough to the obtained one experimentally (their discrepancy makes up ~10%).

The second method of value l_{if}^T estimation consists in using of the two following equations [3, 29]:

$$\varphi_{if} = \varphi_n\left(d_{surf} - 2\right) \tag{7}$$

and

$$\varphi_{if} = \varphi_n\left[\left(\frac{R_p + l_{if}^T}{R_p}\right)^3 - 1\right], \tag{8}$$

where φ_{if} and φ_n are relative volume fractions of interfacial regions and nanofiller, accordingly.

The combination of the indicated equations allows receiving the following formula for l_{if}^T calculation:

$$l_{if}^T = R_p\left[\left(d_{surf} - 1\right)^{1/3} - 1\right] \tag{9}$$

The calculation according to the Equation (9) gives for the considered nanocomposite l_{if}^T = 10.8 nm, that also corresponds well enough to the experiment (in this case discrepancy between l_{if} and l_{if}^T makes up ~ 19%).

Let us note in conclusion the important experimental observation, which follows from the processed by program SPIP results of the studied nanocomposite surface scan (Figure 1). As one can see, at one nanoshungite particle surface from one to three (in average – two) steps can be observed, structurally identified as interfacial layers. It is significant that these steps width (or l_{if}) is approximately equal to the first (the closest to nanoparticle surface) step width. Therefore, the indicated observation supposes, that in elastomeric nanocomposites at average two interfacial layers are formed: the first at the expense of nanofiller particle surface with elastomeric matrix interaction, as a result of which molecular mobility in this layer is frozen and its state is glassy-like one, and the second at the expense of glassy interfacial layer with elastomeric polymer matrix interaction. The most important question from the practical point of view whether one interfacial layer or both serve as nanocomposite reinforcing element. Let us fulfil the following quantitative estimation for this question solution. The reinforcement degree (E_n/E_m) of polymer nanocomposites is given by the equation [3]:

$$\frac{E_n}{E_m} = 1 + 11\left(\varphi_n + \varphi_{if}\right)^{1.7} \tag{10}$$

where E_n and E_m are elasticity moduli of nanocomposite and matrix polymer, accordingly (E_m = 1.82 MPa [3]).

According to the Equation (7) the sum ($\varphi_n + \varphi_{if}$) is equal to:

$$\varphi_n + \varphi_{if} = \varphi_n\left(d_{surf} - 1\right) \tag{11}$$

if one interfacial layer (the closest to nanoshungite surface) is a reinforcing element and

$$\varphi_n + 2\varphi_{if} = \varphi_n\left(2d_{surf} - 3\right) \tag{12}$$

if both interfacial layers are a reinforcing element.

In its turn, the value φ_n is determined according to the equation [30]:

$$\varphi_n = \frac{W_n}{\rho_n} \tag{13}$$

where W_n is nanofiller mass content, ρ_n is its density, determined according to the Equation (6).

The calculation according to the Equations (11) and (12) gave the following E_n/E_m values: 4.60 and 6.65, respectively. Since the experimental value E_n/E_m = 6.10 is closer to the value, calculated according to the Equation (12), then this means that both interfacial layers are a reinforcing element for the studied nanocomposites. Therefore,, the coefficient 2 should be introduced in the equations for value l_{if} determination (for example, in the Equation (1)) in case of nanocomposites with elastomeric matrix. Let us remind that the Equation (1) in its initial form was obtained as a relationship with proportionality sign that is without fixed proportionality coefficient [21].

Thus, the used nanoscopic methodic allow estimating both interfacial layer structural special features in polymer nanocomposites and its sizes and properties. For the first time it has been shown that in elastomeric particulate-filled nanocomposites two consecutive interfacial layers are formed which are a reinforcing element for the indicated nanocomposites. The proposed theoretical methodics of interfacial layer thickness estimation, elaborated within the frameworks of fractal analysis, give well enough correspondence to the experiment.

For theoretical treatment of nanofiller particles aggregate growth processes and final sizes traditional irreversible aggregation models are inapplicable, since it is obvious, that in nanocomposites aggregates a large number of simultaneous growth takes place. Therefore, the model of multiple growths offered [6], was used for nanofiller aggregation description.

In Figure 2 the images of the studied nanocomposites, obtained in the force modulation regime and corresponding to them nanoparticles aggregates fractal dimension d_f distributions are adduced. As it follows from the adduced values d_f^{ag} (d_f^{ag} = 2.40–2.48), nanofiller particles aggregates in the studied nanocomposites are formed by a mechanism particle-cluster (P-Cl), so they are Witten-Sander clusters [32]. The variant A, was chosen which according to mobile particles are added to the lattice, consisting of a large number of "seeds" with density of c_0 at simulation beginning [31]. Such model generates the structures, which have fractal geometry on length short scales with value $d_f \approx 2.5$ (see Figure 2) and homogeneous structure on length large scales. A relatively high particles concentration c is required in the model for uninterrupted network formation [31].

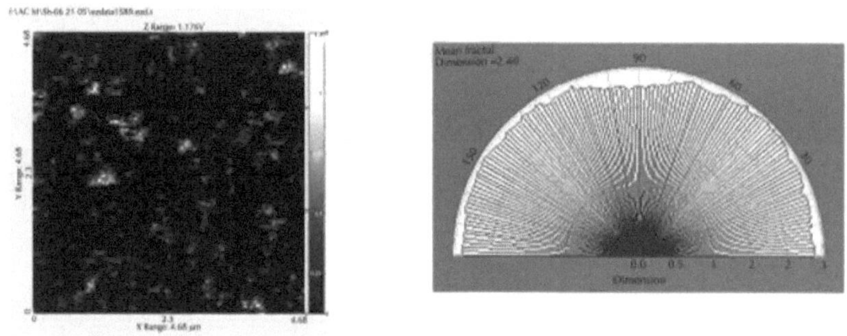

a)

Mean fractal dimension d_f^{ag}=2,40

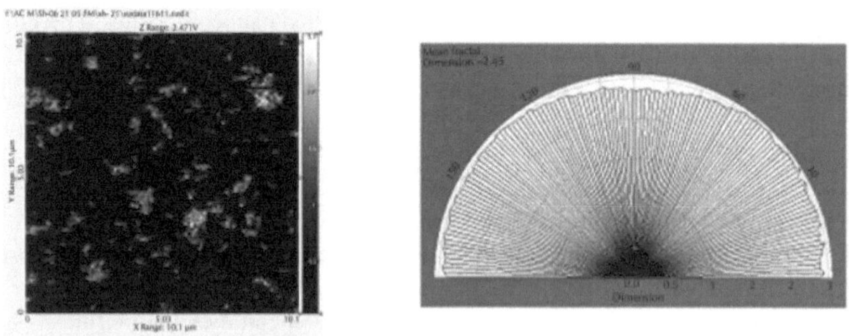

b)

Mean fractal dimension d_f^{ag}=2,45

FIGURE 2 *(Continued)*

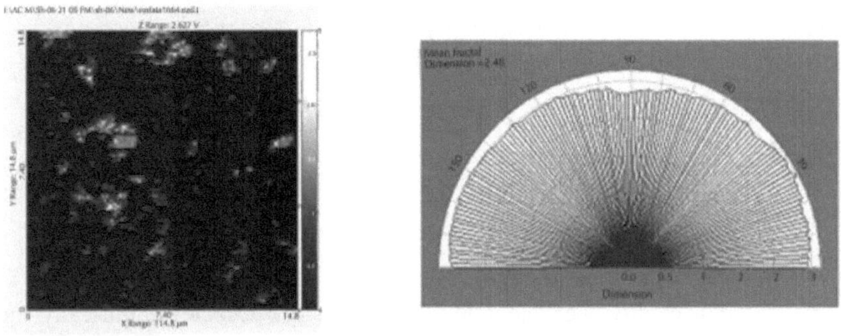

c) Mean fractal dimension d_f^{ag}=2,48

FIGURE 2 The images, obtained in the force modulation regime, for nanocomposites, filled with technical carbon (a), nanoshungite (b), microshungite (c), and corresponding to them fractal dimensions.

In case of "seeds" high concentration c_0 for the variant A the following relationship was obtained [31]:

$$R_{\max}^{d_f^g} = N = c / c_0 \tag{14}$$

where $R_{\max}$ is nanoparticles cluster (aggregate) greatest radius, N is nanoparticles number per one aggregate, c is nanoparticles concentration, c_0 is "seeds" number, which is equal to nanoparticles clusters (aggregates) number.

The value N can be estimated according to the following equation [8]:

$$2R_{\max} = \left(\frac{S_n N}{\pi \eta} \right)^{1/2} \tag{15}$$

where S_n is cross-sectional area of nanoparticles, of which an aggregate consists, η is a packing coefficient, equal to 0.74 [28].

The experimentally obtained nanoparticles aggregate diameter $2R_{ag}$ was accepted as $2R_{\max}$ (Table 1) and the value S_n was also calculated according to the experimental values of nanoparticles radius r_n (Table 1). In Table 1 the values N for the studied nanofillers, obtained according to the indicated method, were adduced. It is significant that the value N is a maximum one for nanoshungite despite larger values r_n in comparison with technical carbon.

Further the Equation (14) allows estimating the greatest radius $R_{\max}^T$ of nanoparticles aggregate within the frameworks of the aggregation model [31]. These values $R_{\max}^T$ are adduced in Table 1, from which their reduction in a sequence of technical carbon-nanoshungite-microshungite, that fully contradicts to the experimental data, that is to R_{ag} change (Table 1). However, we must not neglect the fact that the Equation

(14) was obtained within the frameworks of computer simulation, where the initial aggregating particles size are the same in all cases [31]. For real nanocomposites the values r_n can be distinguished essentially (Table 1). It is expected, that the value R_{ag} or $R^T_{\max}$ will be the higher, the larger is the radius of nanoparticles, forming aggregate is r_n. Then theoretical value of nanofiller particles cluster (aggregate) radius R^T_{ag} can be determined as follows:

$$R^T_{ag} = k_n r_n N^{1/d_f^{ag}} \tag{16}$$

where k_n is proportionality coefficient, in the present work accepted empirically equal to 0.9.

TABLE 1 The parameters of irreversible aggregation model of nanofiller particles aggregates growth.

Nanofiller	R_{ag}, nm	r_n, nm	N	$R^T_{\max}$, nm	R^T_{ag}, nm	R_c, nm
Technical carbon	34.6	10	35.4	34.7	34.7	33.9
Nanoshungite	83.6	20	51.8	45.0	90.0	71.0
Microshungite	117.1	100	4.1	15.8	158.0	255.0

The comparison of experimental R_{ag} and calculated according to the Equation (16) R^T_{ag} values of the studied nanofillers particles aggregates radius shows their good correspondence (the average discrepancy of R_{ag} and R^T_{ag} makes up 11.4 %). Therefore, the theoretical model [31] gives a good correspondence to the experiment only in case of consideration of aggregating particles real characteristics and in the first place, their size.

Let us consider two more important aspects of nanofiller particles aggregation within the frameworks of the model [31]. Some features of the indicated process are defined by nanoparticles diffusion at nanocomposites processing. Specifically, length scale, connected with diffusible nanoparticle, is correlation length ξ of diffusion. By definition, the growth phenomena in sites, remote more than ξ, are statistically independent. Such definition allows connecting the value ξ with the mean distance between nanofiller particles aggregates L_n. The value ξ can be calculated according to the equation [31]:

$$\xi^2 \approx \tilde{n}^{-1} R_{ag}^{d_f^{ag}-d+2} \tag{17}$$

where c is nanoparticles concentration, which should be accepted equal to nanofiller volume contents φ_n, which is calculated according to the Equations (6) and (13).

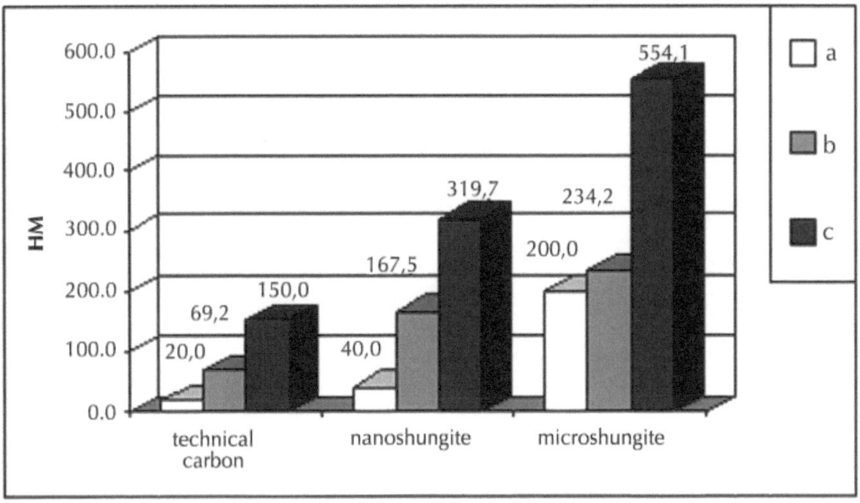

FIGURE 3 The initial particles diameter (a), their aggregates size in nanocomposite (b), and distance between nanoparticles aggregates (c) for nanocomposites, filled with technical carbon, nano and microshungite.

The values r_n and R_{ag} were obtained experimentally (see histogram of Figure 3). In Figure 4 the relation between L_n and ξ is adduced, which, as it is expected, proves to be linear and passing through coordinates origin. This means, that the distance between nanofiller particles aggregates is limited by mean displacement of statistical walks, by which nanoparticles are simulated. The relationship between L_n and ξ can be expressed analytically as follows:

$$L_n \approx 9.6\xi \text{ nm}. \tag{18}$$

The second important aspect of the model [31] in reference to nanofiller particles aggregation simulation is a finite nonzero initial particles concentration c or φ_n effect, which takes place in any real systems. This effect is realized at the condition ξ R_{ag}, that occurs at the critical value $R_{ag}(R_c)$, determined according to the equation [31]:

$$c \sim R_c^{d_f^{ag} - d} \tag{19}$$

The Equation (19) right side represents cluster (particles aggregate) mean density. This equation establishes that fractal growth continues only, until cluster density reduces up to medium density, in which it grows. The calculated according to the Equation (19) values R_c for the considered nanoparticles are adduced in Table 1, from which follows, that they give reasonable correspondence with this parameter experimental values R_{ag} (the average discrepancy of R_c and R_{ag} makes up 24%).

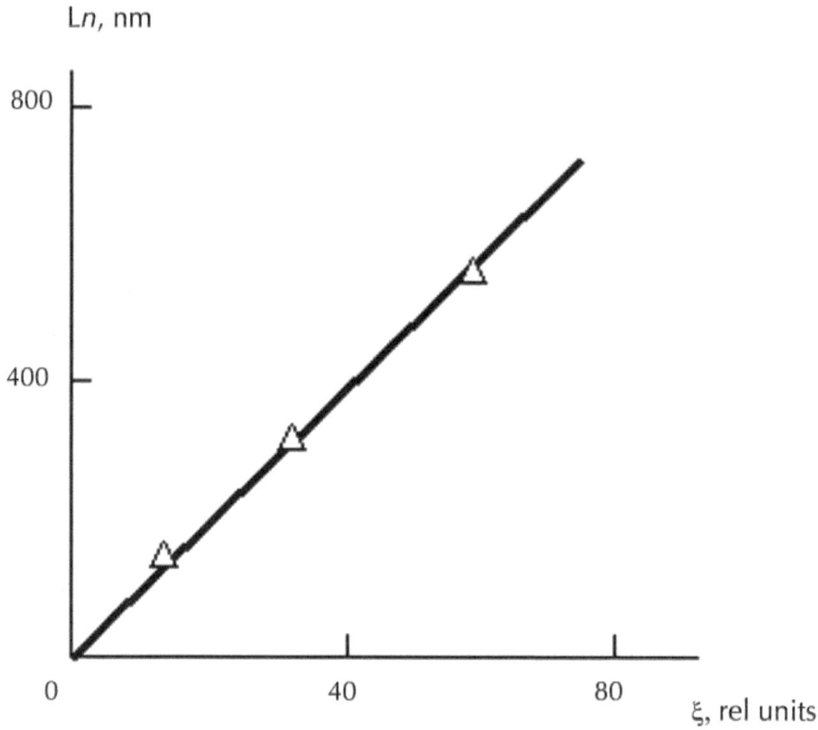

FIGURE 4 The relation between diffusion correlation length ξ and distance between nanoparticles aggregates L_n for considered nanocomposites.

Since the treatment [31] was obtained within the frameworks of a more general model of diffusion-limited aggregation, then its correspondence to the experimental data indicated unequivocally that aggregation processes in these systems were controlled by diffusion. Therefore, let us consider briefly nanofiller particles diffusion. Statistical Walkers diffusion constant ζ can be determined with the aid of the equation [31]:

$$\xi \approx (\zeta t)^{1/2} \tag{20}$$

where t is walk duration.

The Equation (20) supposes (at t = const) ζ increase in a number technical carbon-nanoshungite-microshungite as 196-1069-3434 relative units, that is diffusion intensification at diffusible particles size growth. At the same time diffusivity D for these particles can be described by the well-known Einstein's equation [33]:

$$D = \frac{kT}{6\pi\eta r_n \alpha} \tag{21}$$

where k is Boltzmann constant, T is temperature, η is medium viscosity, α is numerical coefficient, which further is accepted equal to 1.

In its turn, the value η can be estimated according to the equation [34]:

$$\frac{\eta}{\eta_0} = 1 + \frac{2.5\varphi_n}{1-\varphi_n} \tag{22}$$

where η_0 and η are initial polymer and its mixture with nanofiller viscosity, accordingly.

The calculation according to the Equations (21) and (22) shows, that within the indicated nanofillers number the value D changes as 1.32–1.14–0.44 relative units, that is reduces in three times that was expected. This apparent contradiction is due to the choice of the condition t = const (where t is nanocomposite production duration) in the Equation (20). In real conditions the value t is restricted by nanoparticle contact with growing aggregate and then instead of t the value t/c_0 should be used, where c_0 is the seeds concentration, determined according to the Equation (14). In this case the value ζ for the indicated nanofillers changes as 0.288–0.118–0.086, that is it reduces in 3.3 times, that corresponds fully to the calculation according to the Einstein's equation (the Equation (21)). This means, that nanoparticles diffusion in polymer matrix obeys classical laws of Newtonian rheology [33].

Thus, the disperse nanofiller particles aggregation in elastomeric matrix can be described theoretically within the frameworks of a modified model of irreversible aggregation particle-cluster. The obligatory consideration of nanofiller initial particles size is a feature of the indicated model application to real systems description. The indicated particles diffusion in polymer matrix obeys classical laws of Newtonian liquids hydrodynamics. The offered approach allows predicting nanoparticles aggregates final parameters as a function of the initial particles size, their contents, and other factors number.

At present there are several methods of filler structure (distribution) determination in polymer matrix, both experimental [10, 35] and theoretical [4]. All the indicated methods describe this distribution by fractal dimension D_n of filler particles network. However, correct determination of any object fractal (Hausdorff) dimension includes three obligatory conditions. The first from them is the indicated determination of fractal dimension numerical magnitude, which should not be equal to object topological dimension. As it is known [36], any real (physical) fractal possesses fractal properties within a certain scales range. Therefore, the second condition is the evidence of object self-similarity in this scales range [37]. And at the last, the third condition is the correct choice of measurement scales range itself. As it has been shown [38, 39], the minimum range should exceed at any rate one self-similarity iteration.

The first method of dimension D_n experimental determination uses the following fractal relationship [40, 41]:

$$D_n = \frac{\ln N}{\ln \rho} \tag{23}$$

where N is a number of particles with size ρ.

Particles sizes were established on the basis of atomic power microscopy data (see Figure 2). For each from the three studied nanocomposites no less than 200 particles were measured, the sizes of which were united into 10 groups and mean values N and ρ were obtained. The dependences N(ρ) in double logarithmic coordinates were plotted, which proved to be linear and the values D_n were calculated according to their slope (see Figure 5). It is obvious that at such approach fractal dimension D_n is determined in two-dimensional Euclidean space, whereas real nanocomposite should be considered in three-dimensional Euclidean space. The following relationship can be used for D_n recalculation for the case of three-dimensional space [42]:

$$D_3 = \frac{d + D_2 \pm \left[(d - D_2)^2 - 2 \right]^{1/2}}{2} \tag{24}$$

where D_3 and D_2 are corresponding fractal dimensions in three and two-dimensional Euclidean spaces, $d = 3$.

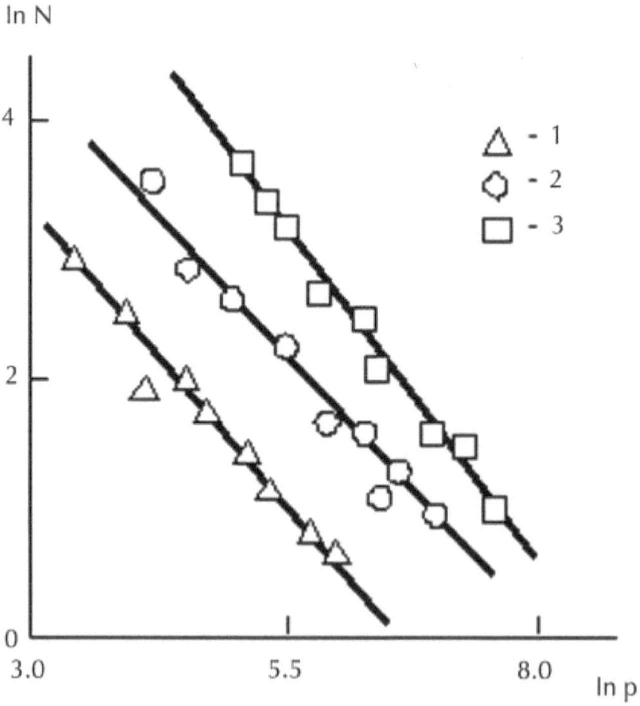

FIGURE 5 The dependences of nanofiller particles number N on their size ρ for nanocomposites BSR/TC (1), BSR/nanoshungite (2), and BSR/microshungite (3).

The calculated according to the indicated method dimensions D_n are adduced in Table 2. As it follows from the data of this table, the values D_n for the studied nanocomposites are varied within the range of 1.10–1.36, so they characterize more or less branched linear formations ("chains") of nanofiller particles (aggregates of particles) in elastomeric nanocomposite structure. Let us remind that for particulate-filled composites polyhydroxiether/graphite the value D_n changes within the range of ~2.30–2.80 [4, 10], that is for these materials filler particles network is a bulk object, but not a linear one [36].

Another method of D_n experimental determination uses the so-called "quadrates method" [43]. Its essence consists in the following. On the enlarged nanocomposite microphotograph (see Figure 2) a net of quadrates with quadrate side size α_i, changing from 4.5 up to 24 mm with constant ratio $\alpha_{i+1}/\alpha_i = 1.5$, is applied and then quadrates number N_i, in to which nanofiller particles hit (fully or partly), is counted up. Five arbitrary net positions concerning microphotograph were chosen for each measurement. If nanofiller particles network is a fractal, then the following equation should be fulfilled [43]:

$$N_i \sim S_i^{-D_n/2} \tag{25}$$

where S_i is quadrate area, which is equal to α_i^2.

In Figure 6 the dependences of N_i on S_i in double logarithmic coordinates for the three studied nanocomposites, corresponding to the Equation (25), is adduced. As one can see, these dependences are linear that allows to determine the value D_n from their slope. The determined according to the Equation (25) values D_n are also adduced in Table 2, from which a good correspondence of dimensions D_n, obtained by the two described methods, follows (their average discrepancy makes up 2.1% after these dimensions recalculation for three-dimensional space according to the Equation (24)).

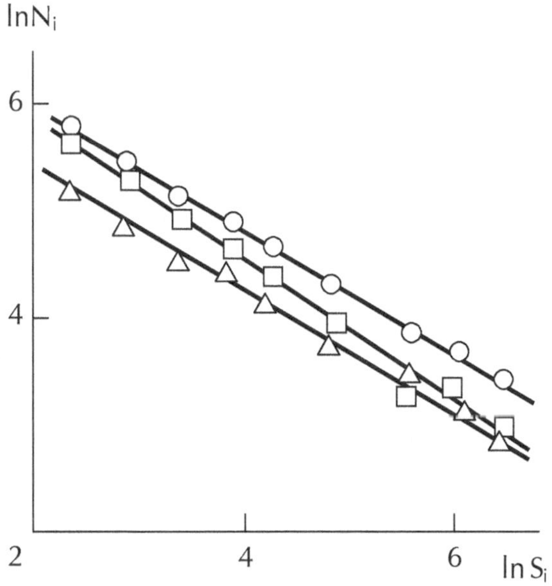

FIGURE 6 The dependences of covering quadrates number N_i on their area S_i, corresponding to the Equation (25), in double logarithmic coordinates for nanocomposites on the basis of BSR. The designations are the same, that in Figure 5.

As it has been shown [44], the usage for self-similar fractal objects at the Equation (25) the condition should be fulfilled:

$$N_i - N_{i-1} \sim S_i^{-D_n} \tag{26}$$

In Figure 7 the dependence, corresponding to the Equation (26), for the three studied elastomeric nanocomposites is adduced. As one can see, this dependence is linear, passes through coordinates origin that according to the Equation (26) is confirmed by nanofiller particles (aggregates of particles) "chains" self-similarity within the selected α_i range. It is obvious that this self-similarity will be a statistical one [44]. Let us note that the points corresponding to $\alpha_i = 16$ mm for nanocomposites butadiene-styrene rubber/technical carbon (BSR/TC) and butadiene-styrene rubber/microshungite (BSR/microshungite), do not correspond to a common straight line. Accounting for electron microphotographs of Figure 2 enlargement this gives the self-similarity range for nanofiller "chains" of 464–1472 nm. For nanocomposite butadiene-styrene rubber/nanoshungite (BSR/nanoshungite), which has no points deviating from a straight line of Figure 7, α_i range makes up 311–1510 nm, that correspond well enough to the indicated self-similarity range.

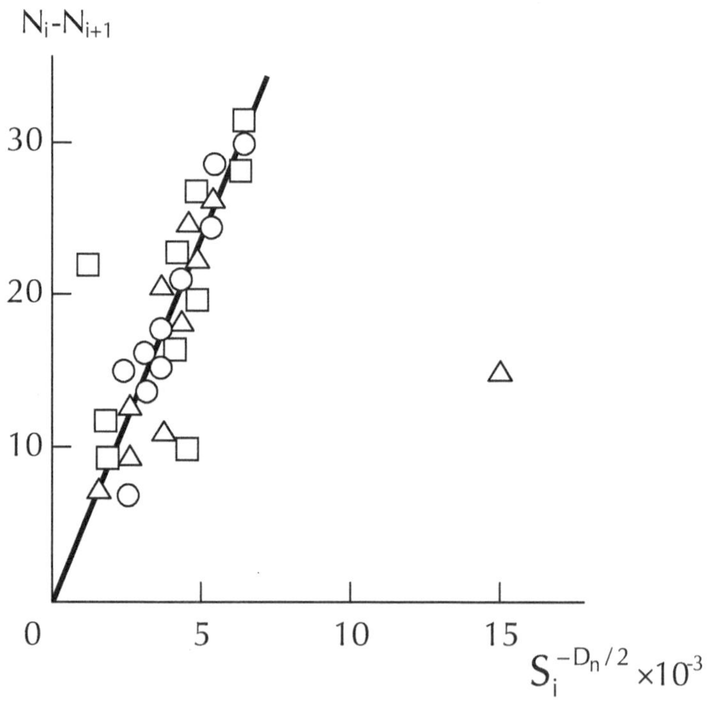

FIGURE 7 The dependences of $(N_i\text{-}N_{i+1})$ on the value $S_i^{-D_n/2}$, corresponding to the Equation (26), for nanocomposites on the basis of BSR. The designations are the same, that in Figure 5.

In observations [38, 39], it has been shown that measurement scales S_i minimum range should contain at least one self-similarity iteration. In this case, the condition for ratio of maximum S_{max} and minimum S_{min} areas of covering quadrates should be fulfilled [39]:

$$\frac{S_{max}}{S_{min}} > 2^{2/D_n} \tag{27}$$

Hence, accounting for the defined restriction let us obtain S_{max}/S_{min} = 121/20.25 = 5.975, that is larger than values $2^{2/D_n}$ for the studied nanocomposites, which are equal to 2.71–3.52. This means, that measurement scales range is chosen correctly.

The self-similarity iterations number μ can be estimated from the inequality [39]:

$$\left(\frac{S_{max}}{S_{min}}\right)^{D_n/2} > 2^{\mu} \tag{28}$$

Using the indicated values of the included in the inequality (28) parameters, μ = 1.42–1.75 is obtained for the studied nanocomposites, that is in the experiment conditions self-similarity iterations number is larger than unity that again confirms correctness of the value D_n estimation [35].

And let us consider in conclusion the physical grounds of smaller values D_n for elastomeric nanocomposites in comparison with polymer microcomposites, that is the causes of nanofiller particles (aggregates of particles) "chains" formation in the first ones. The value D_n can be determined theoretically according to the equation [4]:

$$\varphi_{if} = \frac{D_n + 2.55d_0 - 7.10}{4.18} \tag{29}$$

where φ_{if} is interfacial regions relative fraction, d_0 is nanofiller initial particles surface dimension.

The dimension d_0 estimation can be carried out with the help of the Equation (4) and the value φ_{if} can be calculated according to the Equation (7). The results of dimension D_n theoretical calculation according to the Equation (29) are adduced in Table 2, from which a theory and experiment good correspondence follows. The Equation (29) indicates unequivocally to the cause of filler in nano and microcomposites different behavior. The high (close to 3, see Table 2) values d_0 for nanoparticles and relatively small (d_0 = 2.17 for graphite [4]) values d_0 for microparticles at comparable values φ_{if} is such cause for composites of the indicated classes [3, 4].

Hence, the stated results have shown, that nanofiller particles (aggregates of particles) "chains" in elastomeric nanocomposites are physical fractal within self-similarity (and, hence, fractality [41]) range of ~500–1450 nm. In this range their dimension D_n can be estimated according to the Equations (23), (25), and (29). The cited examples demonstrate the necessity of the measurement scales range correct choice. As it has been noted earlier [45], the linearity of the plots, corresponding to the Equations (23) and (25), and D_n nonintegral value do not guarantee object self-similarity (and, hence, fractality). The nanofiller particles (aggregates of particles) structure low dimensions are due to the initial nanofiller particles surface high fractal dimension.

TABLE 2 The dimensions of nanofiller particles (aggregates of particles) structure in elastomeric nanocomposites.

Nanocomposite	D_n, the Equation (23)	D_n, the Equation (25)	d_0	d_{surf}	φ_n	D_n, the Equation (29)
BSR/TC	1.19	1.17	2.86	2.64	0.48	1.11
BSR/nanoshungite	1.10	1.10	2.81	2.56	0.36	0.78
BSR/microshungite	1.36	1.39	2.41	2.39	0.32	1.47

In Figure 8 the histogram is adduced, which shows elasticity modulus E change, obtained in nanoindentation tests, as a function of load on indenter P or nanoindentation depth h. Since for all the three considered nanocomposites the dependences $E(P)$ or $E(h)$ are identical qualitatively, then further the dependence $E(h)$ for nanocomposite BSR/TC was chosen, which reflects the indicated scale effect quantitative aspect in the most clearest way.

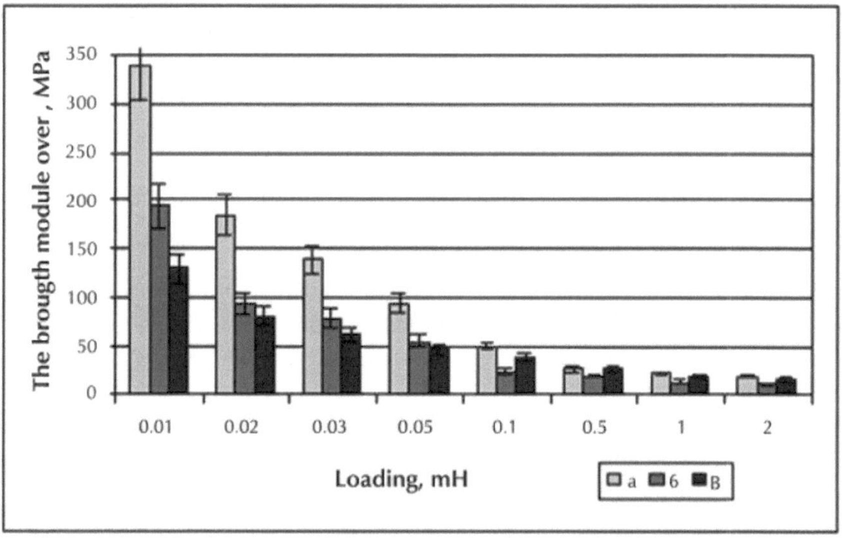

FIGURE 8 The dependences of reduced elasticity modulus on load on indentor for nanocomposites on the basis of butadiene-styrene rubber, filled with technical carbon (a), micro (b), and nanoshungite (c).

In Figure 9 the dependence of E on h_{pl} (see Figure 10) is adduced, which breaks down into two linear parts. Such dependences elasticity modulus strain is typical for polymer materials in general and is due to intermolecular bonds anharmonicity [46]. In study [47], it has been shown that the dependence $E(h_{pl})$ first part at $h_{pl} \leq 500$ nm is not connected with relaxation processes and has a purely elastic origin. The elasticity modulus E on this part changes in proportion to h_{pl} as:

$$E = E_0 + B_0 h_{pl} \tag{30}$$

where E_0 is "initial" modulus, that is modulus, extrapolated to $h_{pl} = 0$, and the coefficient B_0 is a combination of the first and second kind elastic constants. In the considered case $B_0 < 0$. Further Grüneisen parameter γ_L, characterizing intermolecular bonds anharmonicity level, can be determined [47]:

$$\gamma_L \approx -\frac{1}{6} - \frac{1}{2}\frac{B_0}{E_0}\frac{1}{(1-2\nu)}, \tag{31}$$

where ν is Poisson ratio, accepted for elastomeric materials equal to ~0.475 [36].

Calculation according to the Equation (31) has given the following values γ_L:13.6 for the first part and 1.50 for the second one. Let us note the first from γ_L adduced values is typical for intermolecular bonds, whereas the second value γ_L is much closer to the corresponding value of Grüneisen parameter G for intrachain modes [46].

Poisson's ratio ν can be estimated by γ_L (or G) known values according to the formula [46]:

$$\gamma_L = 0.7\left(\frac{1+\nu}{1-2\nu}\right). \tag{32}$$

The estimations according to the Equation (32) gave: for the dependence $E(h_{pl})$ first part ν = 0.462, for the second part ν = 0.216. If for the first part the value ν is close to Poisson's ratio magnitude for nonfilled rubber [36], then in the second part case the additional estimation is required. As it is known [48], a polymer composites (nanocomposites) Poisson's ratio value ν_n can be estimated according to the equation:

$$\frac{1}{\nu_n} - \frac{\varphi_n}{\nu_{TC}} + \frac{1-\varphi_n}{\nu_m} \tag{33}$$

where φ_n is nanofiller volume fraction, ν_{TC} and ν_m are nanofiller (technical carbon) and polymer matrix Poisson's ratio, respectively.

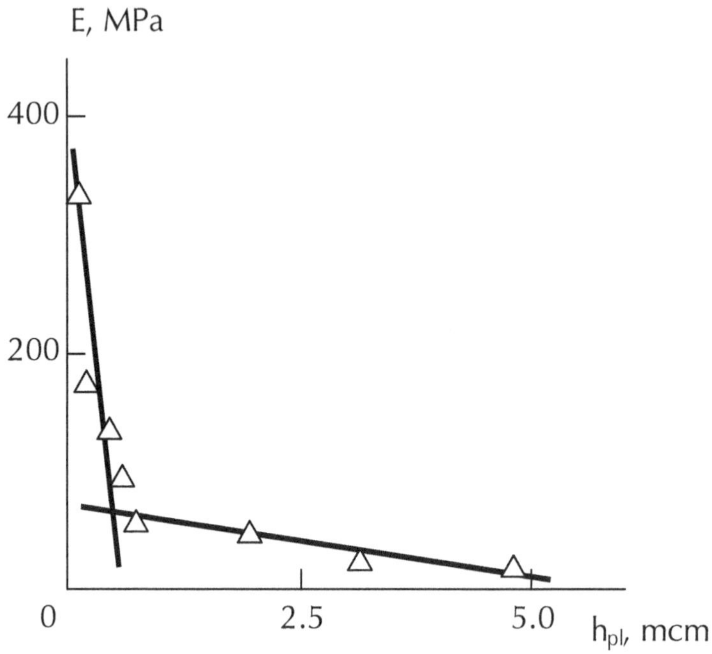

FIGURE 9 The dependence of reduced elasticity modulus E, obtained in nanoindentation experiment on plastic strain h_{pl} for nanocomposites BSR/TC.

The value ν_m is accepted equal to 0.475 [36] and the magnitude ν_{TC} is estimated as follows [49]. As it is known [50], the nanoparticles TC aggregates fractal dimension d_f^{ag} value is equal to 2.40 and then the value ν_{TC} can be determined according to the equation [50]:

$$d_f^{ag} = (d-1)(1+\nu_{TC}). \tag{34}$$

According to the Equation (34) ν_{TC} = 0.20 and calculation ν_n according to the Equation (33) gives the value 0.283, that is close enough to the value ν = 0.216 according to the Equation (32) estimation. As obtained by the indicated methods values ν and ν_n comparison demonstrates, that in the dependence $E(h_{pl})$ (h_{pl}< 0.5 mcm) the first part in nanoindentation tests only rubber-like polymer matrix ($\nu = \nu_m \approx 0.475$) is included and in this dependence the second part – the entire nanocomposite as homogeneous system [51] – $\nu = \nu_n \approx 0.22$.

Let us consider further E reduction at h_{pl} growth (Figure 9) within the frameworks of density fluctuation theory, which value ψ can be estimated as follows [22]:

$$\psi = \frac{\rho_n kT}{K_T} \tag{35}$$

where ρ_n is nanocomposite density, k is Boltzmann constant, T is testing temperature, K_T is isothermal modulus of dilatation, connected with Young's modulus E by the equation [46]:

$$K_T = \frac{E}{3(1-\nu)} \tag{36}$$

In Figure 10 the scheme of volume of the deformed at nanoindentation material V_{def} calculation in case of Berkovich indentor using is adduced and in Figure 11 the dependence $\psi(V_{def})$ in logarithmic coordinates was shown. As it follows from the data of this figure, the density fluctuation growth is observed at the deformed material volume increase. The plot $\psi(\ln V_{def})$ extrapolation to $\psi = 0$ gives $\ln V_{def} \approx 13$ or $V_{def}(V_{def}^{cr}) = 4.42 \times 10^5$ nm^3. Having determined the linear scale l_{cr} of transition to $\psi = 0$ as $(V_{def}^{cr})^{1/3}$, let us obtain $l_{cr} = 75.9$ nm, that is close to nanosystems dimensional range upper boundary (as it was noted, conditional enough [6]), which is equal to 100 nm. Thus, the stated results suppose, that nanosystems are such systems, in which density fluctuations are absent, always taking place in microsystems.

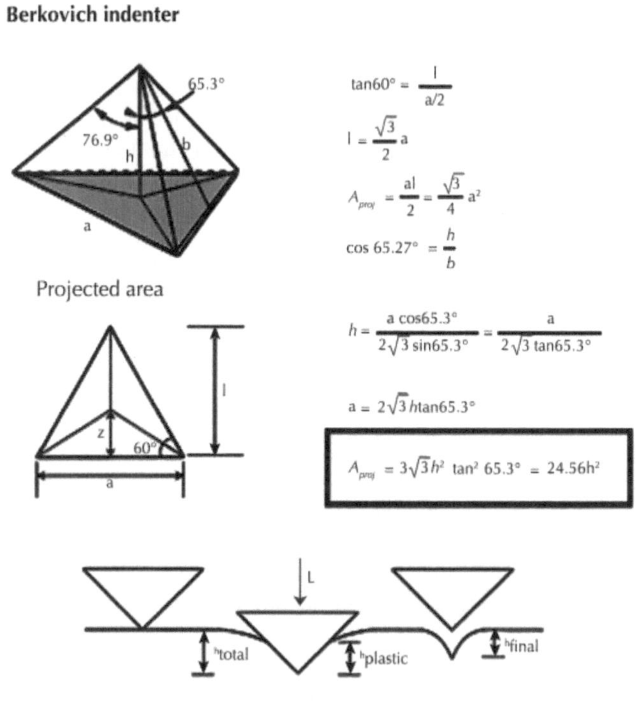

FIGURE 10 The schematic image of Berkovich indentor and nanoindentation process.

As it follows from the data of Figure 9, the transition from nano to microsystems occurs within the range h_{pl} = 408–726 nm. Both the indicated values h_{pl} and the corresponding to them values $(V_{def})^{1/3} \approx$ 814–1440 nm can be chosen as the linear length scale l_n, corresponding to this transition. From the comparison of these values l_n with the distance between nanofiller particles aggregates L_n (L_n = 219.2–788.3 nm for the considered nanocomposites, see Figure 3) it follows, that for transition from nano to microsystems l_n should include at least two nanofiller particles aggregates and surrounding them layers of polymer matrix, that is the lowest linear scale of nanocomposite simulation as a homogeneous system. It is easy to see that nanocomposite structure homogeneity condition is harder than the obtained from the criterion $\psi = 0$. Let us note that such method, namely, a nanofiller particle and surrounding it polymer matrix layers separation is widespread at a relationships derivation in microcomposite models.

It is obvious that the Equation (35) is inapplicable to nanosystems, since $\psi \to 0$ assumes $K_T \to \infty$, that is physically incorrect. Therefore the value E_0, obtained by the dependence $E(h_{pl})$ extrapolation (see Figure 9) to h_{pl} = 0, should be accepted as E for nanosystems [49].

Hence, the stated results have shown that elasticity modulus change at nanoindentation for particulate-filled elastomeric nanocomposites is due to a number of causes which can be elucidated within the frameworks of anharmonicity conception and density fluctuation theory. Application of the first from the indicated conceptions assumes, that in nanocomposites during nanoindentation process local strain is realized, affecting polymer matrix only and the transition to macrosystems means nanocomposite deformation as homogeneous system. The second from the mentioned conceptions has shown that nano and microsystems differ by density fluctuation absence in the first and availability of ones in the second. The last circumstance assumes that for the considered nanocomposites density fluctuations take into account nanofiller and polymer matrix density difference. The transition from nano to microsystems is realized in the case, when the deformed material volume exceeds nanofiller particles aggregate and surrounding it layers of polymer matrix combined volume [49].

In work [3], the following equation was offered for elastomeric nanocomposites reinforcement degree E_n/E_m description:

$$\frac{E_n}{E_m} = 15.2\left[1-\left(d-d_{surf}\right)^{1/t}\right] \qquad (37)$$

where t is index percolation, equal to 1.7 [28].

From the Equation (37) it follows those nanofiller particles (aggregates of particles) surface dimension d_{surf} is the parameter, controlling nanocomposites reinforcement degree [53]. This postulate corresponds to the known principle about numerous division surfaces decisive role in nanomaterials as the basis of their properties change [54]. From the Equations (4)–(6) it follows unequivocally that the value d_{surf} is defined by nanofiller particles (aggregates of particles) size R_p only. In its turn, from the Equation (37) it follows, that elastomeric nanocomposites reinforcement degree E_n/E_m is defined by the dimension d_{surf} only, or accounting for the said, by the size R_p only. This means that the reinforcement effect is controlled by nanofiller particles (aggregates of particles) sizes only and in virtue of this is the true nanoeffect.

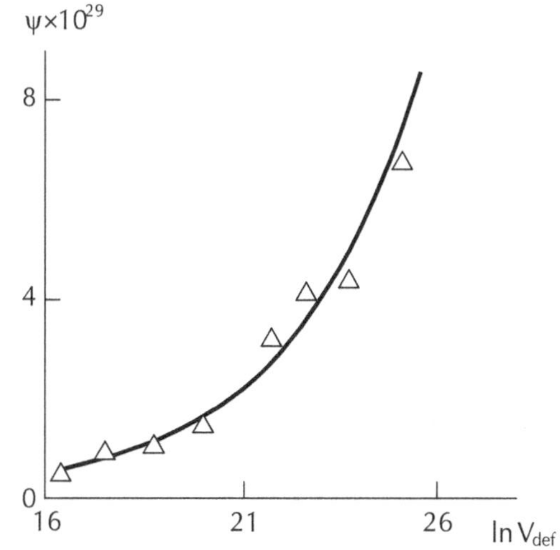

FIGURE 11 The dependence of density fluctuation ψ on volume of deformed in nanoindentation process material V_{def} in logarithmic coordinates for nanocomposites BSR/TC.

In Figure 12 the dependence of E_n/E_m on $(d\text{-}d_{surf})^{1/1.7}$ is adduced, corresponding to the Equation (37), for nanocomposites with different elastomeric matrices (natural and butadiene-styrene rubbers (NR and BSR), accordingly) and different nanofillers (technical carbon of different marks, nano and microshungite). Despite the indicated distinctions in composition, all adduced data are described well by the Equation (37).

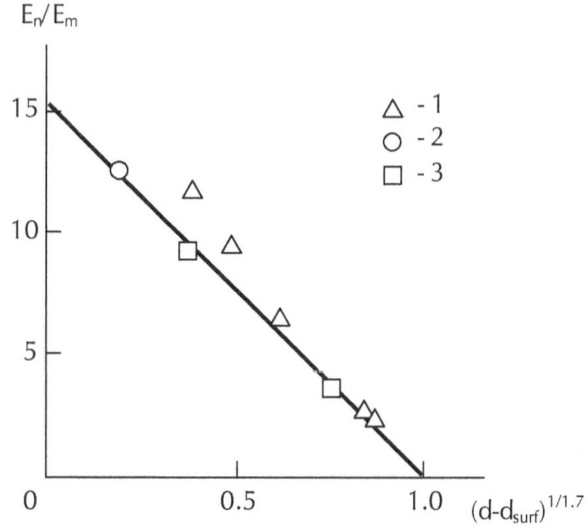

FIGURE 12 The dependence of reinforcement degree E_n/E_m on parameter $(d\text{-}d_{surf})^{1/1.7}$ value for nanocomposites NR/TC (1), BSR/TC (2) and BSR/shungite (3).

In Figure 13 two theoretical dependences of E_n/E_m on nanofiller particles size (diameter D_p), calculated according to the Equations (4)–(6) and (37) are adduced. However, at the curve 1 calculation the value D_p for the initial nanofiller particles was used and at the curve 2 calculation – nanofiller particles aggregates size D_p^{ag} (see Figure 3). As it was expected [5], the growth E_n/E_m at D_p or D_p^{ag} reduction, in addition the calculation with D_p (nonaggregated nanofiller) using gives higher E_n/E_m values in comparison with the aggregated one (D_p^{ag} using). At $D_p \leq 50$ nm faster growth E_n/E_m at D_p reduction is observed than at $D_p >50$ nm, that was also expected. In Figure 13 the critical theoretical value D_p^{cr} for this transition, calculated according to the indicated general principles [54] is pointed out by a vertical shaded line. In conformity with these principles the nanoparticles size in nanocomposite is determined according to the condition, when division surface fraction in the entire nanomaterials volume makes up about 50% and more. This fraction is estimated approximately by the ratio $3l_{if}/D_p$, where l_{if} is interfacial layer thickness. As it was noted, the data of Figure 1 gave the average experimental value $l_{if} \approx 8.7$ nm. Further from the condition $3l_{if}/D_p \approx 0.5$ let us obtain $D_p \approx 52$ nm, that is shown in Figure 13 by a vertical shaded line. As it was expected, the value $D_p \approx 52$ nm is a boundary one for regions of slow ($D_p >52$ nm) and fast ($D_p \leq$ 52 nm) E_n/E_m growth at D_p reduction. In other words, the materials with nanofiller particles size $D_p \leq 52$ nm ("super reinforcing" filler according to the terminology [5]) should be considered true nanocomposites.

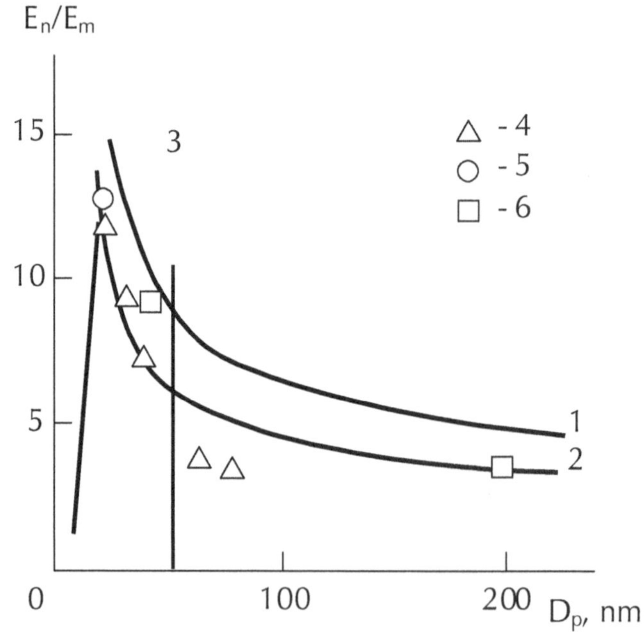

FIGURE 13 The theoretical dependences of reinforcement degree E_n/E_m on nanofiller particles size D_p, calculated according to the Equations (4)–(6), and (37), at initial nanoparticles (1) and nanoparticles aggregates (2) size using. (3) the boundary value D_p, corresponding to true nanocomposite. Equations (4)–(6) – the experimental data for nanocomposites NR/TC (4), BSR/TC (5), and BSR/shungite (6).

Let us note in conclusion that although the curves 1 and 2 of Figure 13 are similar ones, nanofiller particles aggregation which the curve 2 accounts for, reduces essentially enough nanocomposites reinforcement degree. At the same time the experimental data correspond exactly to the curve 2, that was to be expected in virtue of aggregation processes which always took place in real composites [4] (nanocomposites [55]). The values d_{surf}, obtained according to the Equations (4)–(6), correspond well to the determined experimentally ones. So, for nanoshungite and two marks of technical carbon the calculation by the indicated method gives the following d_{surf} values: 2.81, 2.78, and 2.73, whereas experimental values of this parameter are equal to: 2.81, 2.77, and 2.73,that is practically a full correspondence of theory and experiment was obtained.

4.4 CONCLUSION

Hence, the stated results have shown that the elastomeric reinforcement effect is the true nanoeffect which is defined by the initial nanofiller particles size only. The indicated particles aggregation, always taking place in real materials, changes reinforcement degree quantitatively only, namely, reduces it. This effect theoretical treatment can be received within the frameworks of fractal analysis. For the considered nanocomposites the nanoparticle size upper limiting value makes up ~52 nm.

KEYWORDS

- **Butadiene-styrene rubber**
- **Linear spatial structure**
- **Polymer materials**
- **Polymer matrix**
- **Quantitative analysis**

REFERENCES

1. Yanovskii, Yu. G., Kozlov, G. V., and Karnet, Yu. N. *Mekhanika Kompozitsionnykh Materialov i Konstruktsii*, **17**(2), 203–208 (2011).
2. Malamatov, A. Kh., Kozlov, G. V., and Mikitaev, M. A. *Reinforcement Mechanisms of Polymer Nanocomposites*. Publishers of the D.I. Mendeleev RKhTU, Moscow, p. 240 (2006).
3. Mikitaev, A. K., Kozlov, G. V., and Zaikov, G. E. *Polymer Nanocomposites Variety of Structural Forms and Applications*. Nauka, Moscow, p. 278 (2009).
4. Kozlov, G. V., Yanovskii, Yu. G., and Karnet, Yu. N. *Structure and Properties of Particulate-Filled Polymer Composites the Fractal Analysis*. Al'yanstransatom, Moscow, p. 363 (2008).
5. Edwards, D. C. *J. Mater. Sci.*, **25**(12), 4175–4185 (1990).
6. Buchachenko, A. L. *Uspekhi Khimii*, **72**(5), 419–437 (2003).
7. Kozlov, G. V., Yanovskii, Yu. G., Burya, A. I., and Aphashagova, Z. Kh. *Mekhanika Kompozitsionnykh Materialov i Konstruktsii*, **13**(4), 479–492 (2007).
8. Lipatov, Yu. S. The Physical Chemistry of Filled Polymers. *Khimiya*, Moscow, p. 304 (1977).
9. Bartenev, G. M. and Zelenev, Yu. V. *Physics and Mechanics of Polymers*. Vysshaya Shkola, Moscow, p. 391 (1983).
10. Kozlov, G. V. and Mikitaev, A. K. *Mekhanika Kompozitsionnykh Materialov i Konstruktsii*, **2**(3–4), 144–157 (1996).

11. Kozlov, G. V., Yanovskii, Yu. G., and Zaikov, G. E. *Structure and Properties of Particulate-Filled Polymer Composites the Fractal Analysis*. Nova Science Publishers Inc., New York, p. 282 (2010).
12. Mikitaev, A. K., Kozlov, G. V., and Zaikov, G. E. *Polymer Nanocomposites Variety of Structural Forms and Applications*. Nova Science Publishers Inc., New York, p. 319 (2008).
13. McClintok, F. A. and Argon, A. S. *Mechanical Behavior of Materials*. Reading, Addison-Wesley Publishing Company Inc., p. 440 (1966).
14. Kozlov, G. V. and Mikitaev, A. K. *Doklady AN SSSR*, **294**(5), 1129–1131 (1987).
15. Honeycombe, R. W. K. *The Plastic Deformation of Metals*. Boston, Edward Arnold (Publishers) Ltd., p. 398 (1968).
16. Dickie, R. A. In *book Polymer Blends. V. 1*. Academic Press, New York, San-Francisco, and London, pp. 386–431 (1980).
17. Kornev, Yu. V., Yumashev, O. B., Zhogin, V. A., Karnet, Yu. N., and Yanovskii, Yu. G. *Kautschuk i Rezina*, (6), 18–23 (2008).
18. Oliver, W. C. and Pharr, G. M. *J. Mater. Res.*, **7**(6), 1564–1583 (1992).
19. Kozlov, G. V., Yanovskii, Yu. G., and Lipatov, Yu. S. *Mekhanika Kompozitsionnykh Materialov i Konstruktsii*, **8**(1), 111–149 (2002).
20. Kozlov, G. V., Burya, A. I., and Lipatov, Yu. S. *Mekhanika Kompozitnykh Materialov*, **42**(6), 797–802 (2006).
21. Hentschel, H. G. E. and Deutch, J. M. *Phys. Rev. A*, **29**(3), 1609–1611 (1984).
22. Kozlov, G. V., Ovcharenko, E. N., and Mikitaev, A. K. *Structure of Polymers Amorphous State*. Publishers of the D.I. Mendeleev RKhTU, Moscow, p. 392 (2009).
23. Yanovskii, Yu. G. and Kozlov, G. V. *Mater. VII Intern. Sci.-Pract. Conf.* New Polymer Composite Materials. Nal'chik, KBSU, pp. 189–194 (2011).
24. Wu, S. J. *Polymer Sci. Part B Polymer Phys.*, **27**(4), 723–741 (1989).
25. Aharoni, S. M. *Macromolecules*, **16**(9), 1722–1728 (1983).
26. Budtov, V. P. *The Physical Chemistry of Polymer Solutions.*, Khimiya, Saint Peterburg, p. 384 (1992).
27. Aharoni, S. M. *Macromolecules*, **18**(12), 2624–2630 (1985).
28. Bobryshev, A. N., Kozomazov, V. N., Babin, L. O., and Solomatov, V. I. *Synergetics of Composite Materials*. NPO ORIUS, Lipetsk, p. 154 (1994).
29. Kozlov, G. V., Yanovskii, Yu. G., and Karnet, Yu. N. *Mekhanika Kompozitsionnykh Materialov i Konstruktsii*, **11**(3), 446–456 (2005).
30. Sheng, N., Boyce, M. C., Parks, D. M., Rutledge, G. C., Abes, J. I., and Cohen, R. E. *Polymer*, **45**(2), 487–506 (2004).
31. Witten, T. A. and Meakin, P. *Phys. Rev. B*, **28**(10), 5632–5642 (1983).
32. Witten, T. A. and Sander, L. M. *Phys. Rev. B*, **27**(9), 5686–5697 (1983).
33. Happel, J. and Brenner, G. *Hydrodynamics at Small Reynolds Numbers*. Mir, Moscow, p. 418 (1976).
34. Mills, N. J. *J. Appl. Polymer Sci.*, **15**(11), 2791–2805 (1971).
35. Kozlov, G. V., Yanovskii, Yu. G., and Mikitaev, A. K. *Mekhanika Kompozitnykh Materialov*, **34**(4), 539–544 (1998).
36. Balankin, A. S. *Synergetics of Deformable Body*. Publishers of Ministry Defence SSSR, Moscow, p. 404, (1991).
37. Hornbogen, E. *Intern. Mater. Res.*, **34**(6), 277–296 (1989).
38. Pfeifer, P. *Appl. Surf. Sci.*, **18**(1), 146–164 (1984).
39. Avnir, D., Farin, D., and Pfeifer, P. *J. Colloid Interface Sci.*, **103**(1), 112–123 (1985).
40. Ishikawa, K. *J. Mater. Sci. Lett.*, **9**(4), 400–402 (1990).
41. Ivanova, V. S., Balankin, A. S., Bunin, I. Zh., and Oksogoev, A. A. *Synergetics and Fractals in Material Science*. Nauka, Moscow, p. 383 (1994).
42. Vstovskii, G. V., Kolmakov, L. G., and Terent'ev, V. E. *Metally*, (4), 164–178 (1993).
43. Hansen, J. P. and Skjeitorp, A. T. *Phys. Rev. B*, **38**(4), 2635–2638 (1988).
44. Pfeifer, P., Avnir, D., and Farin, D. *J. Stat. Phys.*, **36**(5/6), 699–716 (1984)

45. Farin, D., Peleg, S., Yavin, D., and Avnir, D. *Langmuir*, **1**(4), 399–407 (1985).
46. Kozlov, G. V. and Sanditov, D. S. *Anharmonical Effects and Physical-mechanical Properties of Polymers*. Nauka, Novosibirsk, p. 261 (1994).
47. Bessonov, M. I. and Rudakov, A. P. *Vysokomolek. Soed. B*, **13**(7), 509–511 (1971).
48. Kubat, J., Rigdahl, M., and Welander, M. *J. Appl. Polymer Sci.*, **39**(5), 1527–1539 (1990).
49. Yanovskii, Yu. G., Kozlov, G. V., Kornev, Yu. V., Boiko, O. V., and Karnet, Yu. N. *Mekhanika Kompozitsionnykh Materialov i Konstruktsii*, **16**(3), 445–453 (2010).
50. Yanovskii, Yu. G., Kozlov, G. V., and Aloev, V. Z. *Mater. Intern. Sci.-Pract. Conf.* "Modern Problems of APK Innovation Development Theory and Practice". Nal'chik, KBSSKhA, pp. 434–437 (2011).
51. Chow, T. S. *Polymer*, **32**(1), 29–33 (1991).
52. Ahmed, S. and Jones, F. R. *J. Mater. Sci.*, **25**(12), 4933–4942 (1990).
53. Kozlov, G. V., Yanovskii, Yu. G., and Aloev, V. Z. *Mater. Intern. Sci.-Pract. Conf.*, dedicated to FMEP 50th Anniversary. Nal'chik, KBSSKhA, pp. 83–89 (2011).
54. Andrievskii, R. A. *Rossiiskii Khimicheskii Zhurnal*, **46**(5), 50–56 (2002).
55. Kozlov, G. V., Sultonov, N. Zh., Shoranova, L. O., and Mikitaev, A. K. *Naukoemkie Tekhnologii*, **12**(3), 17–22 (2011).

5 Modification of Arabinogalactan with Isonicotinic Acid Hydrazide

R. Kh. Mudarisova, L. A. Badykova, and Yu. B. Monakov

CONTENTS

5.1 INTRODUCTION

The reaction of an arabinogalactan (AG) and its oxidized forms with the antituberculosis preparation isonicotinic acid hydrazide (INAH) was investigated. The chemical composition and certain physicochemical properties of the modified compounds were studied. A correlation was found between the concentration of carboxylic acids in the polysaccharides and the drug content in the modified compounds.

Drug discovery is one of the thrust areas of modern medicinal chemistry. The search for and development of new antituberculosis agents have recently become of interest because of the drug resistance of *mycobacteria* to existing drugs. One promising direction for creating such drugs is the addition of common tuberculostatic to polysaccharides [1-3]. It is known that the polysaccharide AG has a broad spectrum of biological activity [4-5]. However, its tuberculostatic activity has not been reported. Herein the modification of AG and its oxidized forms by the antituberculosis drug INAH and the antituberculosis activity of the resulting compounds are studied.

Modified compounds were prepared by the reaction of INAH with AG and its oxidized high-molecular-weight (AG_{HMW}) and low-molecular-weight (AG_{LMW}) fractions. The preparation and structure of the oxidized AG fractions have been reported [6, 7].

5.2 EXPERIMENTAL

We used AG of molecular weight (M) 40,000, oxidized forms with AG_{HMW} (M = 22,000), and AG_{LMW} (M = 4,000). Modified compounds were synthesized using INAH (analytically pure). The INAH and polysaccharides were reacted in water. The proton magnetic resonance (PMR) spectra in D_2O were recorded on a Bruker AM-300 spectrometer (operating frequency 300 MHz), ^{13}C nuclear magnetic resonance (NMR) spectra, with wide-field proton decoupling on the same spectrometer (operating frequency 75.47 MHz). We used 3–5% solutions of polysaccharides and INAH in D_2O with DSS internal standard. The ^{13}C NMR spectra were recorded at 25 ± 0.5°C with a 15 s delay between pulses. The IR spectra in mineral oil were recorded on a Shimadzu spectrophotometer. Optical density was determined on a Specord M-40 instrument. The pH values of solutions were measured using an Anion 4100 pH meter and were adjusted by adding NaOH solution (0.1M). Specific rotation was measured using a Perkin-Elmer Model 141 polarimeter.

5.2.1 General Method for Preparing Modified Compounds

The polysaccharide (1 g, 5.55 mM) and INAH (0.85 g, 5.55 mM) were dissolved separately in distilled water (20 ml each). The polysaccharide solution was stirred vigorously and treated dropwise with the INAH solution at 25°C. The reaction was performed for 3 hr. The product was isolated by precipitation by ethanol and reprecipitated from water in ethanol. The solid was separated, washed with alcohol three times and with diethyl ether and dried in vacuo.

5.2.2 Tuberculostatic Activity

It was studied by serial dilutions using Lowenstein-Jensen medium to which (before testing) the studied compounds were added (1 μg/ml). The test cultures were human *Mycobacterium tuberculosis*. The culture suspensions were prepared from a bacterial standard (500 million microbes /ml, 5 units). The resulting suspension (0.2 ml) was inoculated into tubes with nutrient medium containing the tested compounds of each dilution and without them (controls). The tubes were incubated for 21 days at 37°C.

The synthesized compounds are light-brown to white powders that are very soluble in water and insoluble in acetone, alcohols, and ether.

The angles of rotation of the products were $[\alpha]_D^{25}$ +11.9° (AG), +8.4° (AG_{HMW}), +5.3° (AG_{LMW}), +14.0° (AG + INAH), +20.0° (AG_{HMW} + INAH), and +39.0° (AG_{LMW} + INAH).

5.3 DISCUSSION AND RESULTS

The modified AG compounds (AG + INAH, AG_{HMW} + INAH, and AG_{LMW} + INAH) were studied by spectral methods.

Electronic spectra of the products had an absorption band with λ_{max} (H_2O) 250 nm in contrast with the electronic spectrum of INAH with an absorption band at 262 nm.

Solutions of the polysaccharides do not absorb in this region. The hypsochromic shift in the electronic spectra is probably related to steric hindrance from the polysaccharide matrix that disrupts the coplanarity of the conjugated chromophore and is indicative of donor-acceptor interaction between the polymer and the AG_{LMW} compound.

The IR spectra exhibited low frequency shifts of absorption maxima in the range 3600-3100 cm^{-1} that correspond to hydroxyl stretching vibrations by 130–140 cm^{-1} and of absorption maxima of C–O ether stretching vibrations in the pyranose and furanose rings at 1200–1100 cm^{-1} by 19–20 cm^{-1}. This may indicate formation of intermolecular H-bonds between INAH and the polysaccharides. Furthermore, the absorption band at 1750 cm^{-1} that is typical of carbonyl vibrations in the range 1500–1750 cm^{-1} weakened. An absorption band of medium strength appeared at 1550 cm^{-1} and was attributed to vibrations of the pyridine ring of INAH.

TABLE 1 Effect of modification conditions of AG, AG_{HMW}, and AG_{LMW} by INAH on drug content in the modified form*.

AG:INAH mole ratio	τ, hr	T, °C	INAH content, mol per mol polymer
AG			
1:1	3 (6, 10)	20	0.01
1:1	24 (48)	20	0.02
1:0.1 (1:0.5)	24	20	0.01
1:2 (1:3)	24	20	0.02
1:1	3	0 (40, 60)	0.01
1:1	3	90	0.02
AG_{HMW}			
1:1 (1:2) (1:3)	4	20	0.12
AG_{LMW}			
1:1 (1:3)	4	20	0.40

*Changes of parameters not affecting INAH yield are shown in parentheses.

The NMR spectroscopy confirmed that modified compounds (polysaccharides and INAH) were formed. The PMR spectrum of mixed AG and INAH contained resonances for aromatic protons that were broadened and transformed. The resonances for C-6 and C-2 appeared as a broad singlet (8.75 ppm) instead of a doublet (8.74 ppm, 6.1 Hz). The resonances for C-3 and C-5 (7.75 ppm, 6.2 Hz) also changed from a doublet to a broad singlet at 7.77 ppm.

The ^{13}C NMR spectra of the reaction products of AG and INAH exhibited shifts of the resonances for C-2/C-6 and C-3/C-5 by 0.1–0.15 ppm and a shift of the signal for C-4 by 0.2 ppm. The greatest shift to weak field by 0.3 ppm was seen for C-7. Therefore, the biopolymers were modified at the INAH amino group and the carboxylic acid of AG and its oxidized fractions. Thus, the most probable reaction is coordination of the polysaccharide and INAH as follows:

$$AG–COOH + H_2N–C_6H_5N_2O \rightarrow AG–COOH\cdots H_2N–C_6H_5N_2O \rightarrow AG–COO^{-}\cdots{}^{+}H_3N–C_6H_5N_2O,$$

where AG–COOH = AG, and $H_2N–C_6H_5N_2O$ = INAH.

We studied the effect of the modification conditions of AG, AG_{HMW}, and AG_{LMW} by INAH on the drug content of the modified form (Table 1). The best results were obtained using equimolar amounts of the starting compounds. Table 1 show that increasing the molar ratio by 10 times (from 0.1 to 1.0) increased slightly the amount of bound drug. Then, increasing the amount of added INAH did not increase the INAH content in the products. The reaction temperature and time had no effect on the drug content in the complex. Judging from the results, the reaction was practically complete after 1 hr.

Changing the polysaccharide had a large effect on the composition of the products. The INAH content in them increased on going from AG to its oxidized forms. The greatest INAH content was observed in modified compounds of AG_{LMW}.

This was due to the increased concentration of carboxylic acids in the oxidized AG. Initial AG contained 0.047 mol of COOH groups per mole of polymer; AG_{HMW}, 0.12 and AG_{LMW}, 0.76.

Tuberculostatic tests found that AG and modified compounds of AG, AG_{HMW}, and AG_{LMW} possess tuberculostatic activity *in vitro* against pathogenic mycobacteria of the same level as free INAH. Partial growth inhibition of *mycobacteria tuberculosis* was noted for AG_{HMW} and AG_{LMW}.

5.4 CONCLUSION

Thus, it can be concluded that the INAH amino group and AG carboxylic acids react during modification of AG and its oxidized fractions by INAH. A correlation was found between the concentration of carboxylic acids in the polysaccharides and the drug content in the modified compounds. Modified compounds containing from 1 to 18% of the drug and possessing good tuberculostatic activity were obtained.

KEYWORDS

- **Arabinogalactan**
- **Antituberculosis activity**
- **Chemical modification**
- **Polysaccharides**

REFERENCES

1. Kayukova, L. A. and Praliev, K. D. *Khim.-farm. Zh.*, **34**(1), 12 (2000).
2. Khomyakov, K. P., Virnik, A. D., Ushakov, S. N., and Rogovin, Z. A., *Vysokomol. Soedin.*, **7**(6), 1035 (1965).

3. Dol'berg, E. B., Yasnitskii, B. G., Shuteeva, L. N., and Kovalev, I. P., *Zh. Prikl. Khim.*, **46**(9), 2121 (1973).
4. Ovodov, Yu. S. *Bioorg. Khim.*, **24**(7), 483 (1998).
5. Arifkhodzhaev, A. O. *Khim. Prir. Soedin.*, (3), 185 (2000).
6. Borisov, I. M., Shirokova, E. N., Mudarisova, R. Kh., Muslukhov, R. R., Zimin, Yu. S., Medvedeva, S. A., Tolstikov, G. A., and Monakov, Yu. B. *Izv. Ross. Akad. Nauk, Ser. Khim.*, (2), 305 (2004).
7. Borisov, I. M., Shirokova, E. N., Babkin, V. A., Tolstikov, G. A., and Monakov, Yu. B. *Dokl. Ross. Akad. Nauk*, **383**, 774 (2002).

6 Natural Polymer Sorbents for Wastewater Purification from Industrial Dyes

F. F. Niyazi, V. S. Maltseva, and A. V. Sazonova

CONTENTS

6.1 INTRODUCTION

This chapter shows the comparative characteristics of the sorption properties of polymer cellulose and natural carbonate sorbents. The influence of the mass of sorbents on the degree of extraction, as well as the pH changing is analyzed. The optimum phase ratio has been determined. Kinetic curves have been plotted.

It is almost impossible to find a source of pure drinking water because of man's impact on the environment nowadays. Therefore the hygienic requirements for the composition of water discharged into basins after the contact with human life activity are very high.

The main impurities of water to be purified are ions of metals, nitrogen compounds in the form of amines and nitrates, various organic and inorganic substances, which give water color, turbidity, odor, and so on.

Nowadays one of the most acute problems is the environmental pollution by xenobiotics from industrial wastewater, for example textile mills.

Wastewaters of dyeing and finishing enterprises are multicomponent mixtures with a constantly changing qualitative and quantitative composition. They may contain up to 150 types of water pollutants of mineral and organic origin, among which are the dyes being in the largest amounts. Most modern dyes are artificial and contain nitrogen [1].

6.2 EXPERIMENTAL

The use of organic dyes makes it possible to obtain a vast range of consumer products. In textile industry direct, sulfur, reactive, and vat dyes are commonly used for dyeing. Cationic dyes have appeared recently. They have got their name because like basic dyes, and in contrast to all other water-soluble dyes they dissociate into colored cations and colorless anions [2].

Coming into water bodies, dyes attack communities of aquatic organisms. Most organic dyes are highly toxic and have carcinogenic, mutagenic, and allergic effects. Textile industry wastewater purification is a serious problem. Sorption methods of deep water purification are highly effective in solving this complex problem.

Sorbents used in industrial wastewater purification should possess high sorptive capacity, and have developed surface and high kinetic characteristics. They must be available and affordable. Long-term studies confirm that mineral sorbents (activated carbon, metal oxides, waste production), synthetic materials (resins and fibers), and so on can satisfy these requirements [3].

The polymer sorbents are a promising new category of sorbents. According to their characteristics they are not as good conventional sorbents such as activated carbon. Their advantages are low price, selective properties, and usability. For example, cellulose is a linear polymer.

Industrial production of chain polymers began at the beginning of the 20th century although the background for this emerged earlier. Large tonnage production is based on cellulose. Celluloid, the first polymeric material of the physically modified cellulose, was obtained at the beginning of the 20th century. Large-scale production of cellulose ethers and esters was organized before and after World War II and is carried out up to days [4].

Researches, being conducted in recent years have shown that much attention is given to the use of natural polymer materials. These materials are both of anthropogenic origin and agricultural waste products of vegetable origin, for example, sawdust as a sorbent for destaining from aqueous media [5].

Sawdust is small wood cuttings obtained when sawing. It is a waste product of woodworking industry and its use as a sorbent serves the purpose of improving the ecological situation in the region and creating waste-free and resource saving technologies.

Researchers are taking interest in studying the sorption properties of natural inorganic rocks to compare sorptive power and a number of other indices at purifying wastewaters of dyeing and finishing enterprises using organic polymer and inorganic carbonate sorbents.

To determine the optimal amount of sorbent the dependence of the degree of extraction of the cationic blue dye from aqueous solutions on the mass of the sorbents has been studied. The sorbents are cellulose sorbent (sawdust) and natural mineral (carbonate rocks, chalk deposits of Konyshovka settlement, Kursk region).

The sorbents under study are multicomponent systems and their sorptive properties largely depend on physicochemical composition and properties.

According to the accepted classification the length of sawdust does not exceed 50 mm (larger particles are usually regarded as chips). Sawdust contains carbohydrates about 70% (cellulose and hemicellulose) and lignin about 27%. The chemical balance of substances is the following: 50% of carbon, 6% of hydrogen, 44% of oxygen, and about 0.1% of nitrogen.

Since there is still no standard classification of carbonate rocks according to genesis, composition, and structure, the classification of V. N. Kirkinskaya is the easiest and the most convenient for practical use [6]. The analysis of the sorbent has shown that calcite is the larger portion of the system under study. According to the ratio of calcite and dolomite this carbonate rocks can be referred to the limestone dolomite (ω = 50–25%). Structural and microscopic characteristics of the carbonate rocks under study were determined using a polarization interference microscope BIOLAR by the method of uniform field. The studies have shown that the original rocks are made up of trigonal calcite crystals and rhombic dolomite crystals [7].

We used the method of single-stage static sorption. The model aqueous solutions of cationic blue dye concentration of $C_0 = 0.01$ was added to the sorbents under study. The mixture was stirred with a magnetic stirrer. At regular intervals a sample was taken and the content of the dye was analyzed. The final result is the arithmetic mean of five repeated determinations.

6.3 DISCUSSION AND RESULTS

The dye concentration was determined spectrophotometrically. The absorption spectra of the dye are presented in the following coordinates: the optical density (A) and the wavelength (X) on the device SF-26. According to the absorption spectra of the dye the wavelength of maximum light absorbing was selected for cationic blue dye (610 nm). The boundaries of subordination of dye solutions to the fundamental law of light absorbing and Beer law have been determined and the calibration curves have been made.

The degree of extraction (S,%) shows the percentage of the absolute amount of a substance that is collected by the sorbent. This gives a fairly complete picture of the process nature. This characteristic is an important criterion when determining the optimal conditions for the sorption process and is calculated as follows:

$$S = \frac{(C_0 - C_{OCT}) \times 100}{C_0}$$

where Co = initial concentration, g/l

C_{OCT} = residual concentration, g/l

The impact of sorbent mass on the extraction of cationic blue dye has been studied under the following phase ratio: volume V = 50 ml, concentration of the solution Co = 0.01 g/l, contact time of the mixture t = 30 min, and $pH_0 = 3.41$. It is shown in Figure 1.

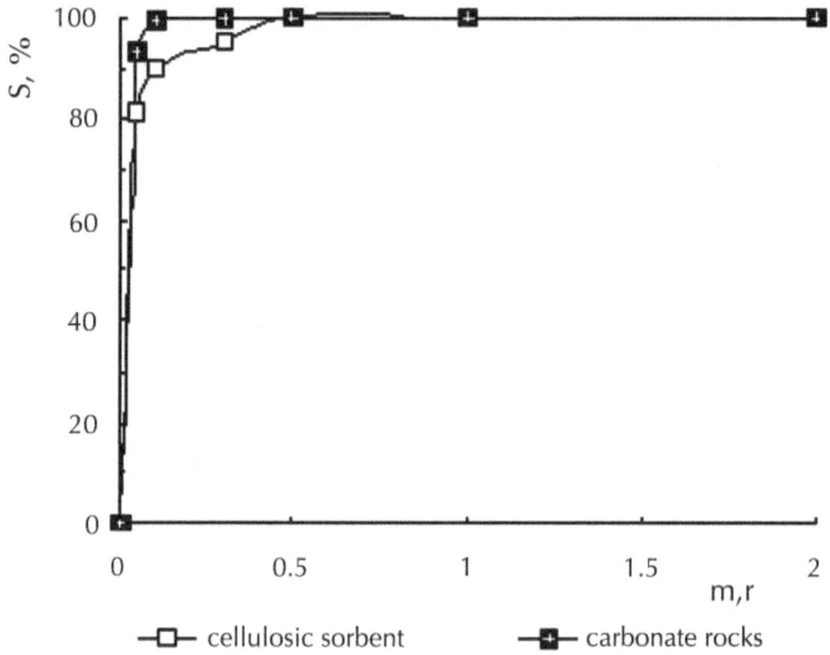

FIGURE 1 The impact of sorbent mass on the extraction of cationic blue dye.

The obtained data show that the sorbents under study efficiently extract cationic blue dye from aqueous solution, sorption is 80–100%. The subsequent increase of the sorbent content does not result in the increase of the degree of the dye extraction. The optimal phase ratio for the systems, in which there is 100% sorption, is determined: "cellulosic sorbent – sorbate volume" as 0.5 g – 50 ml and "carbonate rocks – sorbate volume" as 0.3 g – 50 ml. These data were used in the further work.

The pH value of the environment has a significant influence on the process of sorption and the choice of sorbent for wastewater purification. This is due to the fact that the functional groups located on the surface of the sorbent, and the functional groups of dye molecules may change depending on the pH of the environment [8].

Initial aqueous solution of cationic blue dye had $pH_0 = 3.41$. The dependence of pH on the mass of the studied sorbents after the sorption process is shown in Figure 2.

The obtained dependence shows that pH of the cellulosic sorbent changes slightly with respect to the initial solution. The use of carbonate rocks increases this index to pH 6.5–7.7 depending on the weight. This takes place because of the increased content of hydroxyl groups when hydrolysing carbonates. This is an important characteristic when selecting a sorbent for the neutralization of waste waters in the purification process of industrial acidic drains.

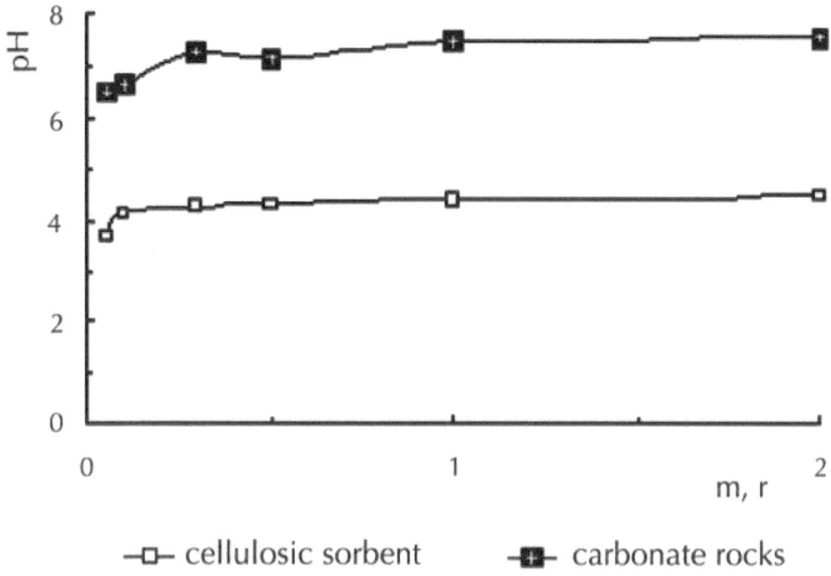

FIGURE 2 The dependence of pH on the mass of the studied sorbents after the sorption process.

This study shows that in the indicated range of pH cationic dye is sorbed on the cellulose sorbent in the form of singly charged cations due to the electrostatic interactions with the oppositely charged functional groups of the sorbent surface.

When studying the characteristics of the sorbents it is necessary to determine the time during which the sorptive equilibrium is established in the system.

The kinetics of the sorption process of cationic blue dye has been studied at the following phase ratio: V = 50 ml, C_0 = 0.01 g/l, and pH_0 = 3.41. The mass of the cellulosic sorbent is 0.5 g, the mass of the carbonate rock is 0.3 g. The results obtained in studying the kinetics of sorption process allow calculating the degree of the dye extraction. In addition, the pH of the medium has been measured. The results are presented in Table 1.

TABLE 1 Kinetics of the sorption process of cationic blue dye by cellulosic sorbent and carbonate rocks.

t, min	**1**	**5**	**10**	**20**	**25**	**30**	**60**
carbonate rocks							
A	0.065	0.035	0.015	0.010	0.005	0	0
Res. conc., $C_{ост}$, g/l	0,0013	0,0008	0,0005	0,0003	0,0001	0	0
S, %	87	92	95	97	99	100	100
pH	7.57	7.60	7.64	7.67	7.44	7.58	7.60

TABLE 1 *(Continued)*

t, min	1	5	10	20	25	30	60
Cellulosic sorbent							
A	0.025	0.015	0.010	0.005	0.0001	0	0
Res. conc., $C_{оcт}$, g/l	0.0006	0.0005	0.0003	0.0001	0.00005	0	0
S, %	94	95	97	99	99.5	100	100
pH	4.05	4.28	4.39	4.18	4.29	4.37	4.40

Based on these results, we have plotted kinetic curves of the sorption of cationic blue dye by the sorbents under study.

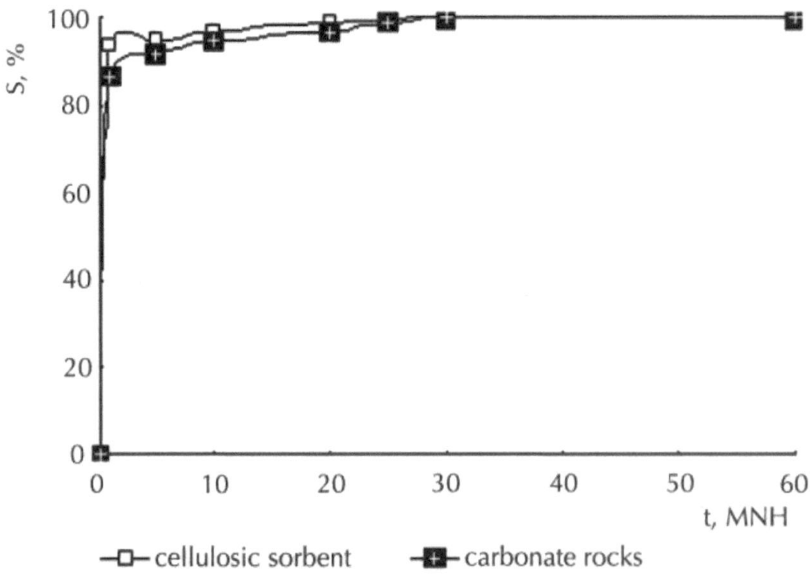

FIGURE 3 Kinetic curves of the sorption of cationic blue dye by the sorbents under study.

The main effect of treatment is achieved in the first 10 min after engaging of the sorbents and the model solution. The time of reaching sorption equilibrium for both cellulosic sorbent and carbonate rocks with the dye is 30 min and further increase of contact time is unreasonable.

When comparing the sorption capacity of the studied sorbents it should be noted that the kinetic curves slightly differ from each other. However, at the same phase ratio and the efficiency of aqueous solutions treatment the carbonate sorbent flow is lower (0.3 g) than cellulose sorbent flow (0.5 g). In addition, carbonate rocks can be recommended as a reagent for neutralization of acidic wastewaters.

It should be noted that it is impractical to regenerate cellulose sorbents, which are waste products of woodworking industry (sawdust) because of their low cost and availability. It is possible to recommend sawdust as a filling in the manufacture of insulation materials and concrete in construction [9-10].

6.4 CONCLUSION

On the basis of these studies it is possible to make an assumption that the sorption of cationic blue dye on carbonate rocks, as well as on cellulosic sorbent is a spontaneous process, but the sorption process is rather complex. The obtained data make it possible to develop affordable, efficient, and environmentally friendly methods of industrial wastewaters treatment.

KEYWORDS

- **Carbonate**
- **Cellulosic sorbent**
- **Degree of extraction**
- **Dye**
- **Kinetics**
- **Polymer**
- **Sorbent**
- **Wastewater purification**

REFERENCES

1. Kruchinina, N. E., Garlenko, M. V., Timasheva, N. A., and Shalbak, A. Decolorization of dyes in textile effluents by electrochemical oxidant. *Safety in tehnosfere*, (1), 10–14 (2009).
2. Sazonova, A. V. Wastewater purification from dye staffs and heavy metal new sorbent. *The Scientific progress - a creative activity young*, Yoshkar-Ola, **1**, 190–191 (2009).
3. Ringqvist, L., Holmgren, A., and Oborn, I. Poorly humified peat as an adsorbent for metals in wastewater. *Water Res*, **36** (9), 2394–2404 (2002).
4. Sperling, L. H. *Introduction to physical polymer science*. Wiley, Hoboken New Jersey p. 10 (2006).
5. Singh, K. K., Rupainwar, D. C., and Hasan, S.H. Low cost bio-sorbent maize bran for the removal of cadmium from wastewater. *J. Indian Chem. Soc.*, **82**(4), 342–346 (2005).
6. Kirkinskaya, V. N. and Laughter, E. M. Karbonatnye sorts-collectors to oils and gas. L.: Depths, p. 255 (1981).
7. Niyazi, F. F., Maltseva, V. S., Burykina, O. V., and Sazonova, A.V. *The Study to sorptions ion honeys from sewages natural carbonate rocks*. The Chemistry in chemistry high school. MGTU N. E. Baumana, p. 151–154 (2010).
8. Bagrovskaya, N. A., Nikiforova, I. E., and Kozlov, V. A. Influence to acidity of the ambience on balance sorption ion Zn(II) and Cd(II) polymer on base of the cellulose. *Journal to general chemistry*, **72**(3), 373–376 (2002).
9. Soldatkina, L. M., Sagaydak, E. V., and Mencuk, V. V. Adsorption cationic dye staffs from water solution on departure of the sunflower. *Chemistry and technology of water*, **31**(4), 417–427 (2009).

10. Cay, S., Uyanik, A., and Ozasik, A. Single and binary component adsorption of copper (II) and cadmium (II) from aqueous solutions using tea-industry waste. *Separ. and Purif. Technol*, **38**(2), 273–280 (2004).

7 The Interaction of Azobisisobutyronitrile with Diketocarboxylic Acid in the Polymerization of Methyl Methacrylate

E. I. Yarmukhamedova, Yu. I. Puzin, and Yu. B. Monakov

CONTENTS

7.1 INTRODUCTION

The influence of aromatic diketocarboxylic acids (DC) on the decomposition initiator of radical polymerization—azobisisobutyronitrile (AIBN) was studied by UV spectroscopy. The interaction occurs with the participation of carboxyl groups of DC with nitrile groups of the initiator. It is shown that polymer obtained in the presence of aromatic DC has mainly a syndiotactic structure.

Among the commonly used initiators of radical polymerization the AIBN is distinguished by the fact that it is inactive in interaction with other substances and induced decomposition. The initiation efficiency of AIBN is only slightly dependent on the nature of the monomer and temperature. Therefore, the increasing of AIBN initiating activity is important.

The number of compounds can effect on the AIBN decomposition. It was established [1] that some mercaptans induce the dissociation of AIBN. Mercaptoacetic and mercaptosuccinic acids, 2-mercaptobenzothiazole decrease the polymerization of methyl methacrylate (MMA) initiated by AIBN: the interaction between the initiator and mercaptan is the reason [2, 3].

Sulfoxide complexes of some transition metals accelerate the polymerization of acrylates initiated by AIBN [4], new complex with azo compounds being formed and its decomposition leads to the initiating of the polymerization. The increase of the polymerization velocity of styrene initiated by AIBN in the presence of 5,6-dihydro-5-methyl-4H-1,3,5-ditiazine (methidine) [5] as well as the number of C-H acids (acetylacetone, fluorene, etc.) [4] was noted. The interaction between phthalic compounds and AIBN was also noted [6], it being primarily proceed with the participation of nitrile groups and the mobile (acidic) hydrogen atoms. It is found that the aromatic DC influence on the kinetic parameters of vinyl monomer polymerization. Therefore, they affect the behavior of azo initiator in the polymerization process. In this regard, the study of the interaction between DC and AIBN is impotent. The structures of DC are shown by Scheme 1.

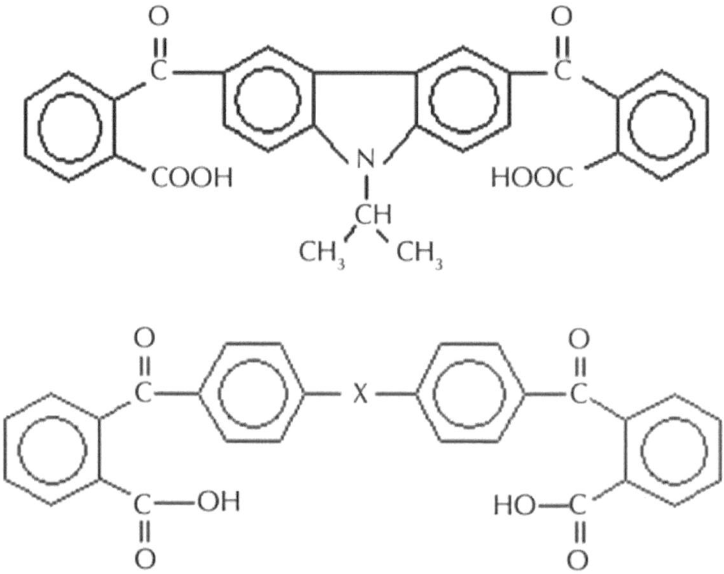

SCHEME 1 3,6-bis(ortho-carboxybenzoyl) - N-isopropylcarbazole (DC-N). 4,4 ′- bis(ortho-carboxybenzoyl) diphenylsulphide (DC-S) and 4,4 ′- bis(ortho-carboxybenzoyl) diphenyloxide (DC-O).

The study of the reaction AIBN with various additives is difficult because the structure of its molecules is symmetrical, weak signals in the infrared (IR) and nuclear magnetic resonance (NMR) are, it does not make decisions about the reaction, especially, at its initial stage. To study the kinetics of the azo initiator decomposition, most often, the spectrophotometric method are used [7].

7.2 EXPERIMENTAL PART

The AIBN was recrystallized from methanol and dried at room temperature under vacuum to constant weight: mp 103°C (with decomposition). Synthesis and purification of heteroaromatic DC was carried out as described [8, 9].

The electronic absorption spectra of solutions of the substances studied were recorded on the spectrophotometer “UV-VIS-NIR 3100” from “Shimadzu”, the temperature was 60°C. The solutions of AIBN and DC in CCl_4 in special sealed quartz cuvettes placed in a thermostated spectrophotometer device. The decomposition of the azo initiator is followed by changes in the absorption band at 345 nm [7]. Spectra of mixtures were recorded in time.

The microstructure of poly(methyl methacrylate) (PMMA) was studied by using NMR ^1H spectroscopy on the spectrophotometer AM-400 from “Bruker” [10]: samples of polymers synthesized in the presence of DC were obtained at the initial stage of polymerization (conversion was 5%) as described [11], was isolated by precipitation from acetone solution in the methanol, dried under vacuum to constant weight at 40°C.

7.3 DISCUSSION AND RESULTS

The interaction between the azo initiator and DC was researched by UV spectroscopic analysis. The DC is low solubility in most solvents and their concentrated solutions (more than 1×10^{-4} mol/l) cannot be practically prepared. The DC concentration was 200 times less than AIBN, and the DC do not have any absorption bands in the area of 345 nm.

After mixing of the azo initiator and DC-N solutions it is observed a hypsochromic effect of the AIBN absorption band (Figure 1) and the band intensity increases.

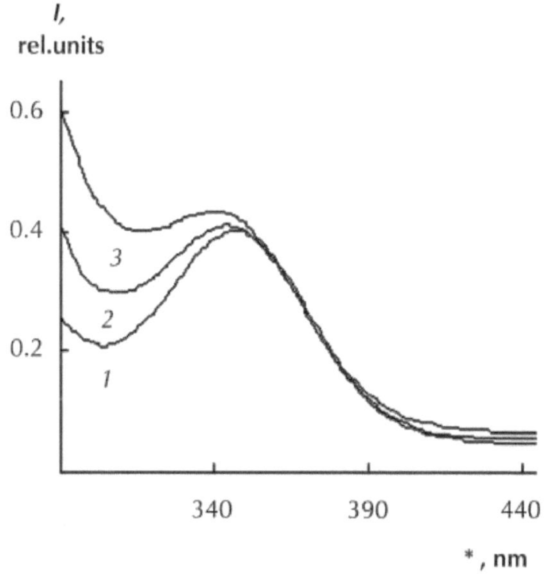

FIGURE 1 Change in the UV spectrum mixture DC - N and AIBN in time: 0 (1), 90 (2), 180 min. (3). Solvent is CCl_4, temperature is 60°C.

The process is accompanied by the appearance and increasing the absorption in the region from 310 to 340 nm (Figure 2). The similar changes are observed in the UV spectra of the mixture of AIBN with DC-O or DC-S to be recorded in the same conditions.

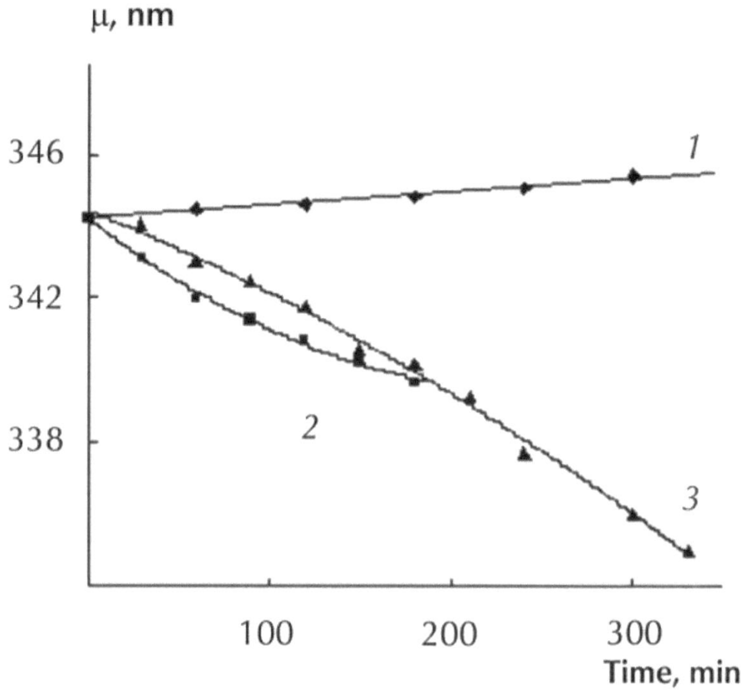

FIGURE 2 Change the position of the absorption band AIBN (1) and in the presence of DC - N (2), DC - O (3), in time.

The analysis of UV spectra of the AIBN solution under similar conditions (Figure 3) showed that the intensity of the band decreases and the peak position at $\lambda = 345$ nm is constant.

Thus, the presence of small amounts of aromatic DC effect on the AIBN decomposition. Similar dependences for DC containing different heteroatoms suggest that the reaction with AIBN involved groups which are in all molecules of the acids studied. It is all carboxyl group. The interaction of cyano groups with acidic compounds, including C-H acids, was noted by several researchers [2, 12]. Probably, the interaction carries out with taking place the carboxyl groups of DC and the nitrile groups of the initiator.

Studying the structure of PMMA obtained in the presence of DC (Table 1), it is obtained the increased content of syndiotactic sequences, even when the synthesis temperature higher 60°C, that is not typical for polymers derived during to free radical polymerization.

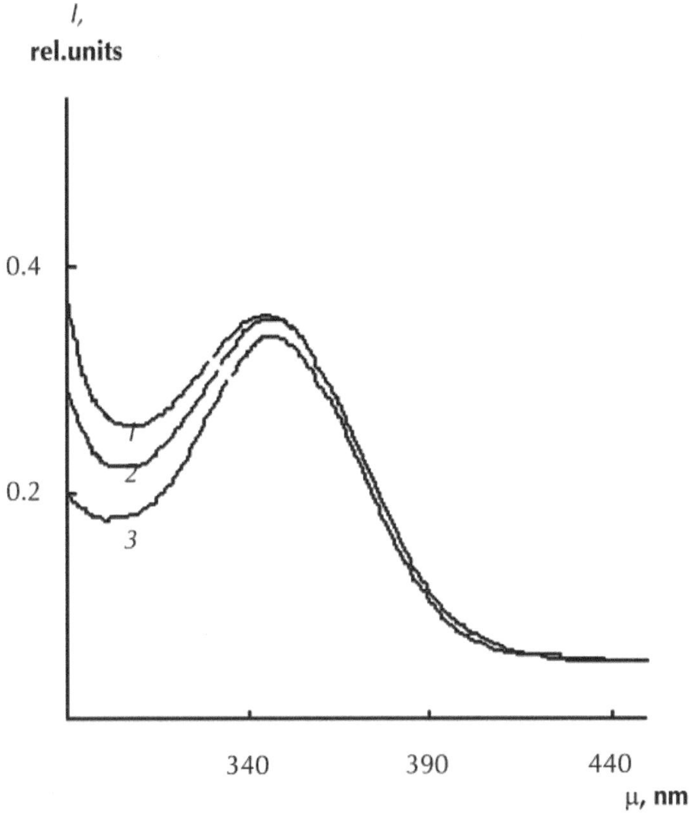

FIGURE 3 Change in the UV range of AIBN solution in time: 0 (1), 120 (2), and 300 min. (3). Solvent is CCl_4, temperature is 60°C.

TABLE 1 Structure of PMMA obtained in the presence of DC. Initiator is 1×10^{-3} mol/l AIBN.

№	Synthesis conditions			Triad content, %		
	[DC], $\times 10^4$ mol /L		T, °C	syndio-	hetero-	iso-
1	DC-N	4,2	60	68	30	2
2	---"---	2,1	60	65	33	2
3	---"---	4,2	75	63	35	2
4	---"---	2,1	75	62	36	2
5	---"---	1,05	75	61,6	36,3	2
6	DC -S	2,5	70	61	36	3
7	---"---	5,0	70	70	28	2
8	DC -O	2,5	60	61	35	4
9	without DC	0	60	56	42	2

7.4 CONCLUSION

Therefore, we can assume that the changes in the microstructure of the polymer are due to the participation in the polymerization not only free radicals but complex linked particulates.

KEYWORDS

- **Azobisisobutyronitrile**
- **Diketocarboxylic acid**
- **Methyl methacrylate**
- **Polymer microstructure**
- **Radical polymerization**

REFERENCES

1. Berger, K. C. and Mahabadi, H. K. IUPAK Macro Mainz: 26th Int. Symp. Macromol. *Mainz.. Short Commun.*, **1**, 83 (1979).
2. Dwivedi, U. N. and Mitra, B. C. *J. Appl. Polym. Sci.*, **29**(12), 4199 (1984).
3. Okada, Y. and Oono, Y. *Kobunsni ronbunshu*, **41**(7), 371 (1984).
4. Puzin, Yu. I. *Doctoral (Chem.) Dissertation.* Ufa: Institute of Organic Chemistry, Ural Scientific Centre, Russian Academy of Sciences (1996).
5. Puzin, Yu. I., Chainikova, E. M., Sigaeva, N. N., Kozlov, V. G., and Leplanin, G. V. *V'isocomolec. Soed.*, **35**(1), 24 (1993).
6. Puzin, Yu. I., Egorov, A. E., and Kraikin, V. A. *Europ. Polym. J.*, **37**, 1167 (2001).
7. Szafko, J., Pabin Szafko, B., Onderko, K., and Wiesniewska, E. *Polimery*, **47**(1), 22 (2002).
8. Zolotukhin, M. G., Egorov, A. E., Sedova, Э. A., and Sorokina, Yl. L. *Doklady Akademii Nauk S.S.S.R.*, **311**(1), 127 (1990).
9. Organikum. Workshop on Organic Chemistry. Izdatel'stvo Mir, **1**. Moscow (1979).
10. Bovey, F. A. *High-resolution NMR of macromolecules*. I. Y. Slonima (Ed.). Izdatel'stvo Chemistry, Moscow (1977).
11. Gladyshev, G. P. *Vinyl monomers polymerization*. Izdatel'stvo Nauka, Alma–Ata (1964).
12. Brutan, E. G. and Fadeev, Yr. A. *Izv. VUSov Physics*, **25**(3), 108 (1982).

8 Radical Polymerization of Methyl Methacrylate in the Presence of Nitrogen Compound

E. I. Yarmukhamedova, Yu. I. Puzin, and Yu. B. Monakov

CONTENTS

8.1 INTRODUCTION

The influence of the 1,3,5-trimethyl-hexahydro-1,3,5-triazine (TMT) on the radical polymerization of methyl methacrylate (MMA) was studied. The kinetic parameters were obtained (reaction orders and activation energy of polymerization). It is established that the triazine is the slight chain transfer during to polymerization initiated by azobisisobutyronitrile (AIBN) and the component of initiating system if the peroxide initiator is used. Polymers synthesized in the presence of TMT have the higher content of sindio and isotactic sequences in macromolecule.

It is known that the organic nitrogen compounds form the initiating systems with peroxide applied for the initiation complex–radical polymerization in nonaqueous media [1, 2]. In this case, the parameters of peroxide decomposition as well as the activity and further reaction of formed radicals change considerably. As a result, the structure of the synthesized polymer is modified.

So, the rate and degree of polymerization change similarly in the presence of 3,6-bis-(ortho-carboxybenzyl)-9-isopropyl-carbazole and the polymers synthesized have the higher amount of syndiotactic sequences [3].

Shown [4] that not only the velocity and degree of polymerization of MMA but also the microstructure of poly(methyl methacrylate) (PMMA) can be regulated in the presence of cyclic hexamethylenetetramine.

Therefore, the investigation of the effect of cyclic triamine TMT on the MMA polymerization and properties of the polymer obtained is interesting. The structure of this compound is shown by scheme 1. In the structure the nitrogen atoms and the geminal methylene groups are: These groups and atoms are most active in formation of redox initiating systems (ROX) for complex–radical polymerization.

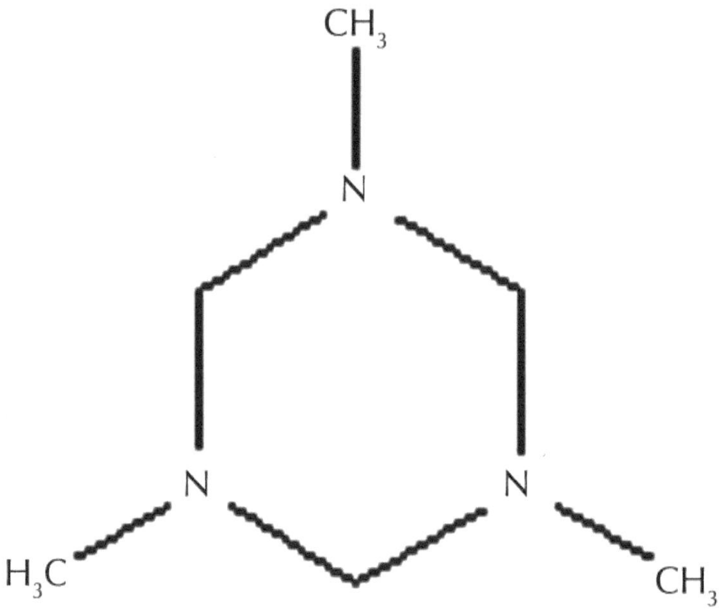

SCHEME 1 The 1,3,5-Trimethyl-hexahydro-1,3,5-triazine (TMT).

8.2 EXPERIMENTAL

The MMA was purified from the stabilizer by double distillation under vacuum. A fraction with a boiling temperature of 42°C/13.3 kPa was used in the polymerization experiments. The AIBN and benzoyl peroxide (BP) were crystallized three times from methanol (melting point 103°C and 108°C with decomposition accordingly) and dried in vacuum at room temperature [5]. The TMT was received by Aldrich (№ 25, 463–3) and was used without purification.

The kinetics of mass polymerization was studied by dilatometery [6]. The polymerization temperature was maintained with an accuracy of ±0.05°C.

The determination of the polymerization order on peroxide or amine is based on the concentration and temperature dependence of the polymerization rate in consequence with the well-known formula [6]:

$$W_0 = k_{eff} \times [M]_0 \times [PB]_0^n \times [Additive]_0^m$$

where k_{eff} is the effective rate constant of polymerization, $[M]_0$ is the initial concentration of monomer, $[PB]_0$, and $[Additive]_0$–initial concentration of initiator or additives, accordingly: n and m–effective polyme rization order on peroxide or triazine, respectively.

Molecular weight characteristics (weight average and number average molecular weights M_w and M_n and molecular weight distribution (MWD) measured as M_w/M_n ratio) of PMMA were determined using gel permeation chromatography on a "Waters GPC 2000 System" liquid chromatograph using tetrahydrofuran as an eluent with flow rate 0.5 ml/min at 30°C. The column system was calibrated using polystyrene standards with narrow MWD ($M_w/M_n < 1.2$), using the universal Benoit dependence and the equation connecting molecular weight of PMMA with characteristic viscosity in tetrahydrofuran.

Determination of the content of syndio and isotactical sequences in PMMA was carried out according [7]. The solvent was $CDCl_3$, the operating frequency of the "Bruker AM-300" NMR spectrometer was 300 MHz, and internal standard was tetramethylsilane (TMS).

8.3 DISCUSSION AND RESULTS

Study of MMA polymerization initiated by the system polybutadiene (PB)—TMT showed (Table 1) that the initial rates as well as the values of M_w and M_n are changed, when the amine concentration increased, in different ways depending on the polymerization temperature. If it is 50°C, they do not practically depend on the additive concentrations. If the polymerization temperature is 60°C, the dependence of the initial velocity *versus* the TMT concentration has a maximum, but that for M_w and M_n values has a minimum. If the temperature is 75°C, the reducing of initial rate and growth of molecular mass of polymer are determined.

TABLE 1 Dependence of the MMA polymerization velocity and molecular mass parameters of polymers obtained *versus* the triazine concentration. Initiator concentrations were $[BP]_0 = 1$ 10^{-3} mol/l.

T, °C	[TMT] ×10³, mol/l	W_0×10³, mol/l ×min	M_w ×10⁻⁵	M_n ×10⁻⁵	MMP
50	0	1,7	34,4	17,2	2,0
	0,5	1,7	32,3	17,0	1,9
	1,0	1,5	34,0	16,9	2,0
	2,0	1,6	35,1	17,5	2,0

TABLE 1 *(Continued)*

T, °C	[TMT] $\times 10^3$, mol/l	$W_0 \times 10^3$, mol/l ×min	M_w $\times 10^{-5}$	M_n $\times 10^{-5}$	MMP
60	0	4,0	30,0	15,1	2,0
	0,5	4,5	24,2	11,4	2,0
	1,0	4,3	23,7	11,5	2,0
	2,0	3,5	23,9	12,6	1,8
75	0	12,3	11,6	5,8	2,0
	0,5	12,1	12,3	6,5	1,9
	1,0	11,6	12,7	6,7	1,9
	2,0	10,5	13,6	6,9	1,9

The polymerization orders on the initiator and amine were calculated. They are 0 on TMT and 0.5 on PB (50°C), 0 on TMT, and 0.4 on PB (60 and 75°C). Since the order on PB is near to 0.5, we can conclude that the polymerization in the presence of TMT occurs by a radical mechanism with quadratic chain termination. At 60 and 75°C the order on the initiator decreases that is probably due to the interaction of system components and the formation of ROS.

Based on the temperature dependences of polymerization rate, the effective activation energy of polymerization in the presence of TMT was calculated to be equal to 45 ± 4 kJ/mol. This value is markedly lower than that in the case of initiation by PB only (80 ± 4 kJ/mol).

The rate of polymerization of MMA at the stage of the gel effect is reduced in the presence of nitrogen-containing compound (Figure 1 and 2).

The increase of TMT concentration in the system during to the polymerization of MMA at 75°C results in the smoothing of gel, effect until its full disappearance at the ratio of the concentrations of additives and initiator, 2:1 (Figure 2). The rate of the process at the stage of the gel effect is probably reduced due to the interaction of amine and PB.

Thus, the triazine slows down the polymerization of MMA initiated by PB at the stage of the gel effect to reduce its velocity.

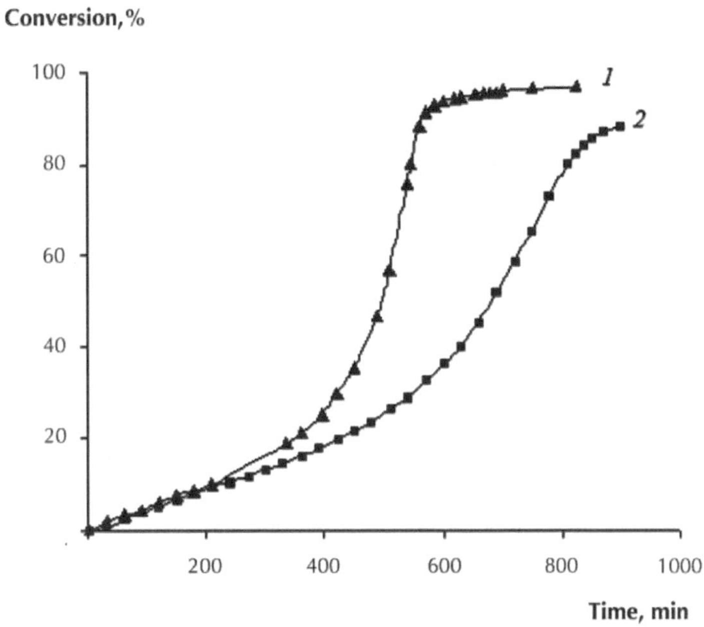

FIGURE 1 Kinetics of the MMA polymerization at 60°C initiated by the [PB] = 1×10^{-3} mol/l, with no additives (1) and in the presence of [TMT] = 2×10^{-3} mol/l (2).

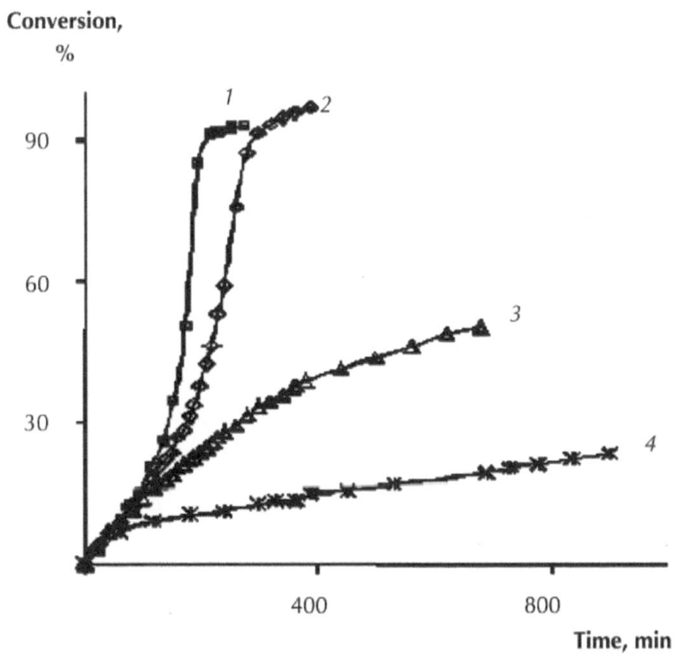

FIGURE 2. Kinetics of MMA polymerization at 75°C initiated by 1×10^{-3} mol/l BP (1) and in the presence of TMT: 1×10^{-3} (2), 2×10^{-3} (3), or 4×10^{-3} (4) mol/l.

The number of organic nitrogen compounds can forms the initiating systems with peroxide, the rate of polymerization in their presence being reduced with increasing of nitrogen compound concentration, most often [8]. Taking into consideration the published data on this reaction, we can assume the following scheme of interaction between components of PB—TMT system (Scheme 2).

It includes the stages of interaction to be typical for the radical initiating systems peroxides organic amines such as the formation of donor-acceptor complex K_1 (charge-transfer complex) which may subsequently transform by radical (formation of K_2) or ion-radical (formation of K_3) ways, and each of them is capable of the initiation of polymerization in the presence of the monomer. In addition, the formation of the K_2 complex takes into account the catalytic function of triazine. The carrying out of several consecutive and parallel processes may be the reason of changing the role of additives which is observed when the polymerization temperature varies.

Obviously, in the initial stages of the polymerization the triazine is a catalyst of BP decomposition which can be caused by the formation of complex between amine and initiator. This, in turn, leads to decrease in the activation energy. However, during to the polymerization the initiator is almost exhausted and the process is slow down.

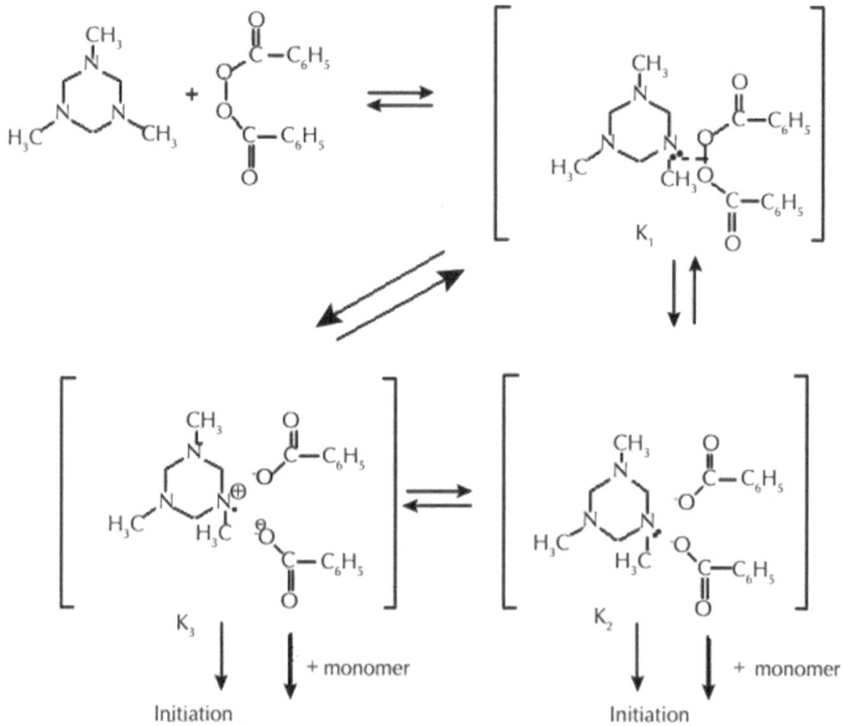

SCHEME 2 Scheme of interaction between components of PB–TMT system.

The study of MMA polymerization initiated by AIBN showed that the addition of nitrogen containing heterocyclic compound has little effect on the initial rate of polymerization and the molecular weight of synthesized polymers (Table 2).

The reaction orders on the initiator and additives were found to be: 0 on TMT and 0.47 on AIBN (50°C), 0 on TMT and 0.57 on AIBN (60°C), and 0 on TMT, and 0.54 on AIBN (75°C). Since the order on the azo initiator is near to 0.5, we can conclude that in this case the polymerization proceeds *via* a radical mechanism with quadratic chain termination in the presence of the amine.

The effective activation energy in the presence of TMT is equal to 89 ± 4 kJ/mol that is slightly higher than that in the case of initiation by AIBN (80 ± 4 kJ/mol) only.

The study of the polymerization to high monomer conversion in the presence of TMT showed that the start of gel effect shifts to a higher conversion (Figure 3).

TABLE 2 Dependence of the MMA polymerization velocity and molecular mass parameters of polymers obtained on the triazine concentration. The initiator concentrations were $[AIBN]_0 = 1 \times 10^{-3}$ mol/l.

T, °C	[TMT] $\times 10^3$, моль/л	$W_0 \times 10^3$, моль/л×мин	M_w $\times 10^{-5}$	M_n $\times 10^{-5}$	MMP
50	0	2,4	26,8	13,2	2,0
	0,5	2,4	26,3	12,9	2,0
	1,0	2,4	26,1	12,7	2,0
	2,0	2,4	25,8	12,6	2,0
60	0	5,8	15,6	7,7	2,0
	0,5	5,6	15,4	7,7	2,0
	1,0	5,8	15,0	7,6	1,9
	2,0	5,9	14,8	7,4	2,0
75	0	19,0	10,6	5,3	2,0
	0,5	18,8	10,3	5,2	2,0
	1,0	18,9	10,0	5,2	1,9
	2,0	19,0	9,9	5,0	2,0

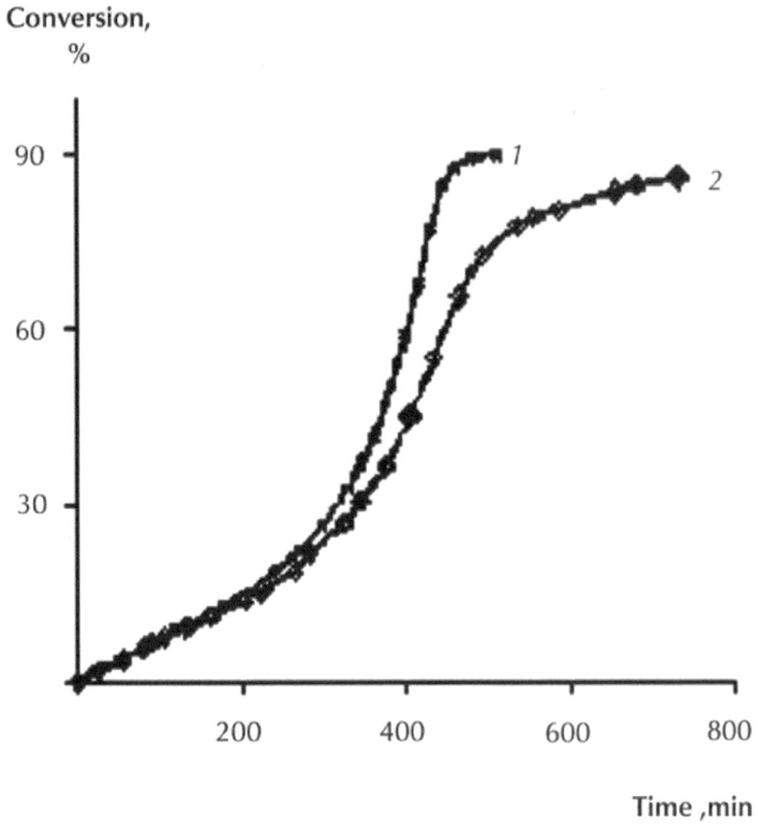

FIGURE 3 Kinetics of MMA polymerization at 60°C initiated by 1×10^{-3} mol/l AIBN (1) and in the presence of 2.0×10^{-3} mol/l TMT (2).

The reducing of the polymerization rate at the stage of gel effect can be result of the chain transfer to the additive or the interaction with the growing macro radicals (Scheme 3).

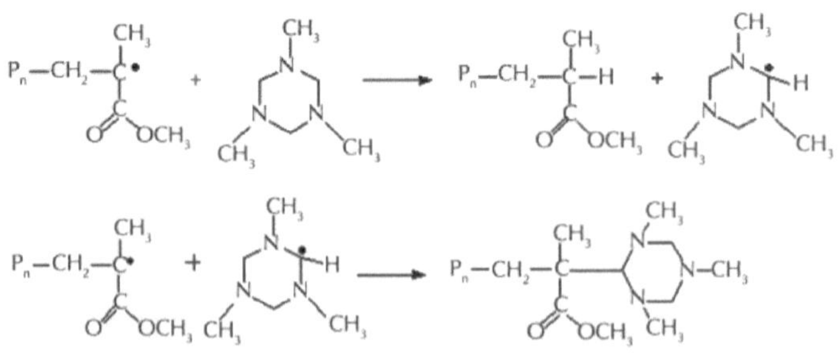

SCHEME 3 Chain transfer to the additive or the interaction with the growing macro radicals.

Studying the microstructure of PMMA obtained in the presence of TMT and initiator AIBN or PB, we found the higher amount of syndiotactic and, most importantly, isotactic sequences (triads) compared with polymers synthesized in its absence (Table 3).

TABLE 3 The microstructure of PMMA prepared in the presence of a triazine derivative [initiator] = 1×10^{-3}mol / l.

Synthesis conditions			Triad content, %		
T, °C	**Initiator**	**[TMT]×10³, моль/л**	**syndio -**	**hetero -**	**iso -**
50	PB	0	60	37	3
		0,50	63	33	4
		1,00	62	34	4
		2,00	62	34	4
	AIBN	0	60	37	3
		0,50	63	33	4
		1,00	62	34	4
		2,00	61	35	4
60	PB	0	56	42	2
		0,25	63	32	5
		0,50	62	33	5
		1,00	61	35	4
		2,00	61	35	4
	AIBN	0	55	43	2
		0,50	62	33	5
		1,00	61	34	5
		2,00	62	33	5

TABLE 3 *(Continued)*

Synthesis conditions			Triad content, %		
T, °C	Initiator	[TMT]×10³, моль/л	syndio -	hetero -	iso -
75	PB	0	53	45	2
		0,25	59	36	5
		0,50	58	37	5
		1,00	59	36	5
		2,00	60	36	4
	AIBN	0	53	45	2
		0,25	59	36	5
		0,50	56	38	6
		1,00	57	37	6
		2,00	57	37	6

Changes in the microstructure and the decrease heterotactic structures, probably are due to the mechanism of MMA polymerization in the presence of triazine is not only free radical but complex radical that.

KEYWORDS

- **Azobisisobutyronitrile**
- **Methyl methacrylate**
- **Molecular weight distribution**
- **Radical polymerization**
- **1,3,5-trimethyl-hexahydro-1,3,5-triazine**

REFERENCES

1. Dolgoplosk, B. A. and Tinyakova, E. I. *The generation of free radicals and their reactions.* Izdatel'stvo Nauka, Moscow (1982).
2. Epimahov, J. K. *V'isokomolek.soedin. B.*, **27**(6), 464 (1985).
3. Puzin, Y. I., Galinurova, E. I., Fatykhov, A. A., and Monakov, J. B. *V'isokomolek.soedin. A.*, **44**(10), 1752 (2002).

4. Yarmukhamedova, E. I., Puzin, Yr. I., and Monakov, Yr. B. *Izv. VUSov. Khimiya i khimitch. tekhnologhiya*, **52**(9), 59 (2009).
5. Toroptseva, A. M., Belgorodskay, K. V., and Bondarenko, V. M. *Laboratory workshop on the chemistry and technology of macromolecular compounds*. Izdatel'stvo Chemistry, Leningrad (1972).
6. Gladyshev, G. P. *Polymerization of vinyl monomers*. Izdatel'stvo Nauka, Alma-Ata (1964).
7. Bovey, F. A. *High-resolution NMR of macromolecules*. I. Y. Slonim. (Ed.). Izdatel'stvo Chemistry, Moscow (1977).
8. Efremova, E. P., Chihaeva, I. P., Stavrova, S. D., Bogachev, Y. S., Zhuravleva, I. L., and Pravednikov, A. N. *V'isokomolek.soedin. A*., **27**(3), 532 (1985).

9 Degradation of Poly(3-Hydroxybutyrate) and its Derivatives: Characterization and Kinetic Behavior

A. P. Bonartsev, A. P. Boskhomodgiev, A. L. Iordanskii, G. A. Bonartseva, A. V. Rebrov, T. K. Makhina, V. L. Myshkina, S. A. Yakovlev, E. A. Filatova, E. A. Ivanov, and D. V. Bagrov

CONTENTS

9.1 INTRODUCTION

The bacterial polyhydroxyalkanoates (PHAs) and their principal representative poly(3-R-hydroxybutyrate) (PHB) create a competitive option to conventional synthetic polymers such as polypropylene, polyethylene, polyesters and so on. These polymers are nontoxic and renewable. Their biotechnology output does not depend on hydrocarbon production as well as their biodegradation intermediates and resulting products (water and carbon dioxide) do not provoke the adverse actions in environmental media or living systems [1-3]. Being friendly environmental [4], the PHB and its derivatives are used as the alternative packaging materials, which are biodegradable in the soil or different humid media [5, 6].

The copolymerization of 3-hydroxybutyrate entities with 3-hydroxyoctanoate (HO) and 3-hydroxyheptanoate (HH) or 3-hydroxyvalerate (HV) monomers modifies the physical and mechanical characteristics of the parent PHB, such as ductility and toughness to depress its processing temperature and embrittlement. Besides, copolymers PHB-HV [7], PHB-HH [8] or PHB-HO [9] and so on have improved thermophysical/mechanical properties and hence they expand the spectrum of constructional and medical materials/items. For predicting the behavior of PHB and its copolymers in an aqueous media for example, *in vitro*, in a living body or in a wet soil, it is essential to study kinetics and mechanism of hydrolytic destruction.

Despite the history of suchlike investigations reckons about 25 years, the problem of biodegradation in semicrystalline biopolymers is too far from a final resolution. Moreover, in the literature the description of hydrolytic degradation kinetics during long term period is comparatively uncommon [10-14]. Therefore, the main object of this chapter is the comparison of long term degradation kinetics for the PLA, PHB, and its derivatives, namely its copolymer with poly(3-hydroxybutyrate-co-3-hydroxyvalerate (PHBV) and the blend PHB/PLA. The contrast between degradation profiles for PHB and PLA makes possible to compare the degradation behavior for two most prevalent biodegradable polymers. Besides, a significant attention is devoted to the impact of molecular weight (MW) for above polymer systems upon hydrolytic degradation and morphology (crystallinity and surface roughness) at physiological (37°C) and elevated (70°C) temperatures.

This work is designed to be an informative source for biodegradable PHB and its derivatives' research. It focuses on hydrolytic degradation kinetics at 37 and 70°C in phosphate buffer to compare PLA and PHB kinetic profiles. Besides, we reveal the kinetic behavior for copolymer PHBV (20% of 3-hydroxyvalerate) and the blend PHB-PLA. The intensity of biopolymer hydrolysis characterized by total weight lost and the viscosity averaged MW decrement. The degradation is enhanced in the series PHBV $<$ PHB $<$ PHB-PLA blend $<$ PLA. Characterization of PHB and PHBV includes MW and crystallinity evolution (X-ray diffraction) as well as atomic force microscopy (AFM) analysis of PHB film surfaces before and after aggressive medium exposition. The important impact of MW on the biopolymer hydrolysis is shown.

9.2 EXPERIMENTAL

9.2.1 Materials

In this work we have used poly(L-lactide) (PLA) with different molecular weights: 67, 152, and 400 kDa (Fluka Germany), chloroform (ZAO EKOS-1, RF), sodium valerate (Sigma-Aldrich, USA), and mono-substituted sodium phosphate (NaH_2PO_4, ChimMed, RF).

9.2.2 PHAs Production

The samples of PHB and copolymer of hydroxybutyrate and hydroxyvalerate (PHBV) have been produced in A. N. Bach's Institue of Biochemistry. A highly efficient strain-producer (80 wt% PHB in the dry weight of cells), *Azotobacter chroococcum* 7Б, has been isolated from rhizosphere of wheat (the sod-podzol soil). Details of PHB biosynthesis have been published in [15]. Under conditions of PHBV synthesis, the sucrose concentration was decreased till 30 g/L in medium and, after 10 hr incubation; 20mM sodium valerate was added. The isolation and purification of the biopolymers were performed *via* centrifugation, washing and drying at 60°C subsequently. Chloroform extraction of PHB or PHBV from the dry biomass and precipitation, filtration, washing again, and drying has been described in work [15]. The monomer content (HB/HV ratio) in PHBV has been determined by nuclear magnetic resonance in accordance with procedure described [16]. The percent concentration of HV moiety in the copolymer was calculated as the ratio between the integral intensity of methyl group of HV (0.89 ppm) and total integral intensity the same group and HB group (1.27 ppm). This value is 21 mol%.

9.2.3 Molecular Weight Determination

The viscosity averaged MW was determined by the viscosity (η) measurement in chloroform solution at 30°C. The calculations of MW have been made in accordance with Mark-Houwink equation [17]:

$$[\eta] = 7.7 \times 10^{-5} \cdot M^{0,82}$$

9.2.4 Film Preparations of PHAs, PLA, and their Blends

The films of parent polymers (PHB, PHBV, and PLA) and their blends with the thickness about 40μm were cast on a fat-free glass surface. We obtained the set of films with different MW = 169 ± 9 (defined as PHB 170), 349 ± 12 (defined as PHB 350), 510 ± 15 kDa (defined as PHB 500), and 950 ± 25 kDa (defined as PHB 1000) as well as the copolymer PHBV with MW = 1056 ± 27 kDa (defined as PHBV). Additionally we prepared the set of films on the base of PLA with same thickness 40μm and MW = 67 (defined as PLA 70), MW = 150 and 400 kDa. Along with them we obtained the blend PHB/PLA with weight ratio 1:1 and MW = 950 kDa for PHB, and MW = 67 kDa for PLA (defined as PHB + PLA blend). Both components mixed and dissolved in common solvent, chloroform and then cast conventionally on the glass plate. All films were thoroughly vacuum-processed for removing of solvent at 40°C.

9.2.5 Hydrolytic Degradation *in vitro* Experiments

The measurement of hydrolytic destruction of the PHB, PLA, PHBV films, and the PHB-PLA composite was performed as follows. The films were incubated in 15 ml 25 mM phosphate buffer, pH 7.4, at 37°C or 70°C in a ES 1/80 thermostat (SPU, Russia) for 91 days; pH was controlled using an Orion 420 + pH meter (Thermo Electron Corporation, USA). For polymer weight measurements films were taken from the buffer solution every 3 day, dried, placed into a thermostat for 1 hr at 40°C and then weighed with a balance. The film samples weighed 50–70 mg each. The loss of polymer weight due to degradation was determined gravimetrically using an AL-64 balance (Acculab, USA). Every 3 days the buffer was replaced by the fresh one.

9.2.6 Wide Angle X-ray Diffraction

The PHB and PHBV chemical structure, the type of crystal lattice and crystallinity was analyzed by wide angle X-ray scattering (WAXS) technique. The WAXS study was performed on device on the basis of 12 kW generator with rotating copper anode RU-200 Rotaflex (Rigaku, Japan) using CuK radiation (wavelength $\lambda = 0.1542$ nm) operated at 40 kV and 140 mA. To obtain pictures of wide angle X-ray diffraction of polymers two-dimensional position sensitive X-ray detector GADDS (Bruker AXS, Germany) with flat graphite monochromator installed on the primary beam was used. Collimator diameter was 0.5 mm [18].

9.2.7 AFM of PHB Films

Microphotographs of the surface of PHB films were obtained be means of AFM. The AFM imaging was performed with Solver PRO-M (Zelenograd, Russia). For AFM imaging a piece of the PHB film (~2×2 mm^2) was fixed on a sample holder by double-side adhesive tape. Silicon cantilevers NSG11 (NT-MDT, Russia) with typical spring constant of 5.1 N/m were used. The images were recorded in semi contact mode, scanning frequency of 1–3 Hz, scanning areas from 3×3 to 20×20 μm^2, topography, and phase signals were captured during each scan. The images were captured with 512×512 pixels. Image processing was carried out using Image Analysis (NT-MDT, Russia) and FemtoScan Online (Advanced technologies center) software.

9.3 DISCUSSION AND RESULTS

The *in vitro* degradation of PHB with different MW and its derivatives (PHBV, blend PHB/PLA) prepared as films was observed by the changes of total weight loss, MW, and morphologies (AFM, XRD) during the period of 91 days.

9.3.1 The Hydrolysis Kinetics of PLA, PHB, and its Derivatives

The hydrolytic degradation of the biopolymer and the derivatives (the copolymer PHBV, and the blend PHB/PLA 1:1) has been monitored for 3 months under condition, which is realistically approximated to physiological conditions, namely, *in vitro*: phosphate buffer, pH = 7.4, and temperature 37°C. The analysis of kinetic curves for all samples shows that the highest rate of weight loss is observed for PLA with the smallest MW ≈ 70 kDa and for PHB with relatively low MW ≈ 150 kDa (Figure 1). On the base of the data in this figure it is possible to compare the weight loss increment for the polymers with different initial MW. Here, we clearly see that the samples with the higher MWs (300–1000 кDa) are much stable against hydrolytic degradation than the samples of the lowest MW. The total weight of PHB films with MW = 150 kDa decreases faster compared to the weight reduction of the other PHB samples with higher MW's = 300 and 450 or 1000 kDa. Additionally, by the 91st day of buffer exposition the residual weight of the low MW sample reaches 10.5% weight loss that it is essentially higher than the weight loss for the other PHB samples (see Figure 1 again).

After establishing the impact of MW upon the hydrolysis, we have compared the weight loss kinetic curves for PLA and PHB films with the relatively comparative MW = 400 and 350 kDa respectively and the same film thickness. For the PLA films one can see the weight depletion with the higher rate than the analogous samples of PHB. The results obtained here are in line with the preceding literature data [8, 12, 19-21].

Having compare destruction behavior of the homopolymer PHB and the copolymer PHBV, we can see that the introduction of hydrophobic entity (HV) into the PHB molecule *via* copolymerization reveals the hydrolytic stability of PHBV molecules. For PHBV a hydrolysis induction time is the longest among the other polymer systems and over a period of 70 days its weight loss is minimal (< 1% wt) and possibly related with desorption of low-molecular fraction of PHBV presented initially in the samples after biosynthesis and isolation. The kinetic curves in Figure.1 show also that the conversion the parent polymers to their blend PHB-PLA decreases the hydrolysis rate compared to PHB (MW = 1000 kDa) even if the second component is a readily hydrolysable polymer: PLA (MW = 70 kDa).

For the sake of hydrolysis amplification and its exploration simultaneously, an polymer exposition in aqueous media has usually been carried out at elevated temperature [11, 19]. To find out a temperature impact on degradation and intensify this process, we have elevated the temperature in phosphate buffer to 70°C. This value of temperature is often used as the standard in other publications, see example [11]. As one should expect, under such condition the hydrolysis acceleration is fairly visible that is presented in Figure 1(b). By the 45th day of PLA incubation its films turned into fine-grinding dust with the weight-loss equaled 50% (MW = 70 kDa) or 40% (MW = 350 kDa). Simultaneously the PHB with the lowest MW = 170 kDa has the weight loss = 38 wt% and the film was markedly fragmented while the PHB samples with higher MWs 350, 500, and 1000 kDa have lost the less percent of the initial weight, namely 20, 15, and 10% respectively. Additionally, for 83 days the weight drop in the PHB-PLA blend films is about 51 wt% and, hence, hydrolytic stability of the blend polymer system is essentially declined (compare Figures 1(a) and 1(b)) .

At elevated temperature of polymer hydrolysis (70°C) as well as at physiological temperature 37°C we have demonstrated again that the PHBV films are the stable because by 95th day they lost only 4 wt%. The enhanced stability of PHBV relative to the PHB has been confirmed by other literature data [21]. Here it is

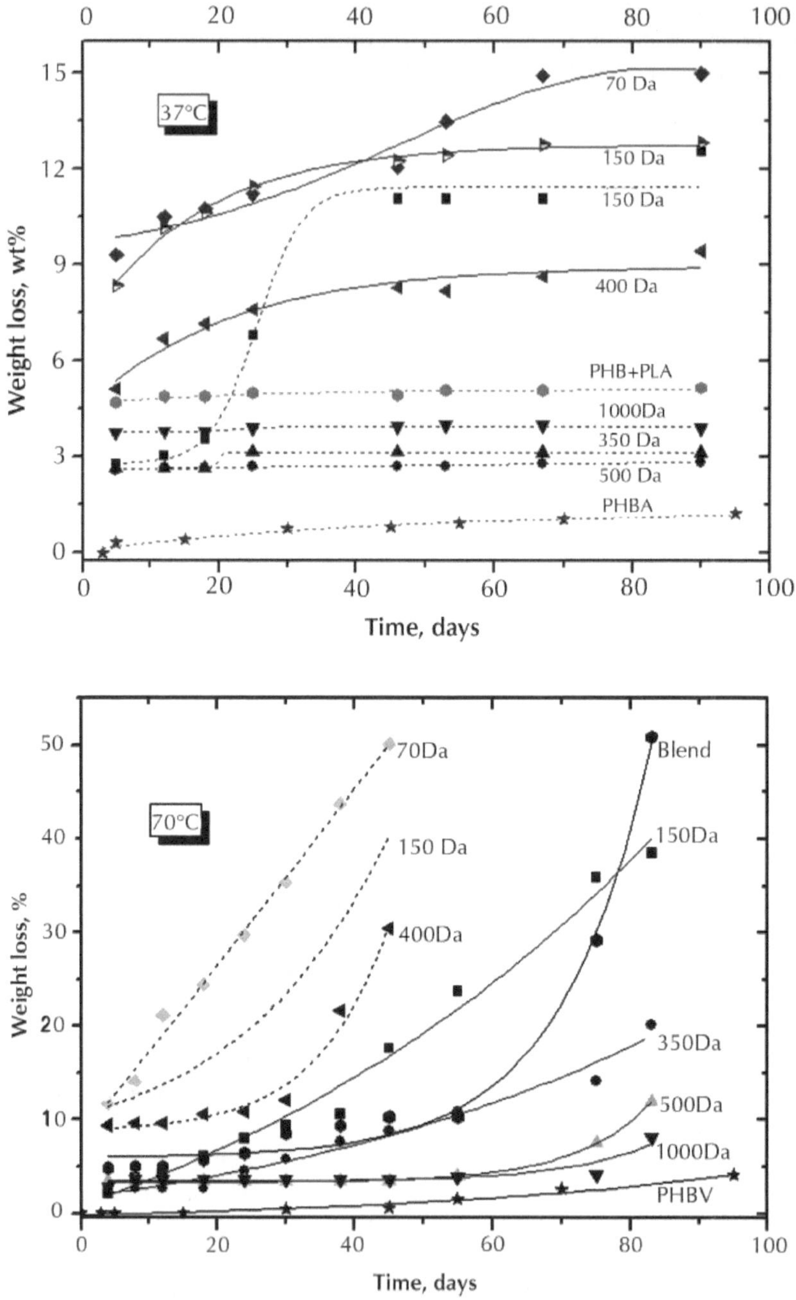

FIGURE 1 Weight loss in the phosphate buffer for PHB and its derivatives with different MW (shown on the curves in kDa). 37°C, 70°C: ♦, ►, and◄ are PLA films with MW = 70, 150, and 400 kDa respectively; ■, ▲, ●, and ▼ are PHB samples with 170, 350, 500, and 1000 kDa, respectively; PHBV 1050 (★); and PHB-PLA blend (⬢).

worth to remark that during biosynthesis of the PHBV two opposite effects of water sorption acting reversely each other occur. On the one side, while the methyl groups are replaced by ethyl groups, the total hydrophobicity of the copolymer is enhanced, on the other side, this replacement leads to decrease of crystallinity in the copolymer [22]. The interplay between two processes determines a total water concentration in the copolymer and hence the rate of hydrolytic degradation. Generally, in the case of PHBV copolymer (HB/HV = 4:1 mol. ratio) the hydrophobization of its chain predominates the effect of crystallinity decrease from 75% for PHB to ~60% for PHBV.

9.3.2 Change of MW for PHB and PHBV

On exposure of PHB and PHBV films to buffer medium at physiological (37°C) or elevated (70°C) temperatures, we have measured both their total weight loss and the change of their MW simultaneously. In particular, we have shown the temperature impact on the MW decrease that will be much clear if we compare the MW decrements for the samples at 37 and 70°C. At 70°C the above biopolymers have a more intensive reduction of MW compared to the reduction at 37°C (see Figure 2). In particular, at elevated temperature the initial MW (= 350 kDa) has the decrement by 7 times more than the MW decrement at physiological condition. Generally, the final MW loss is nearly proportional to the initial MW of sample that is correct especially at 70°C. As an example, after the 83 days incubation of PHB films, the initial MW = 170 kDa dropped as much as 18 wt% and the initial MW = 350 kDa has the 9.1 wt% decrease.

The diagrams in Figure 2 shows that the sharp reduction of MW takes place for the first 45 days of incubation and after this time the MW change becomes slow. Combining the weight-loss and the MW depletion, it is possible to present the biopolymer hydrolysis as the two-stage process. On the initial stage, the random cleavage of macromolecules and the MW decrease without a significant weight-loss occur. Within this time the mean length of PHB intermediates is fairly large and the molar ratio of the terminal hydrophilic groups to the basic functional groups in a biodegradable fragment is too small to provide the solubility in aqueous media. This situation is true for the PHB samples with middle and high MW (350, 500, and 1000 kDa) when at 37°C their total weight remains stable during all time of observation but the MW values are decreased till 76, 61, and 51 wt% respectively. On the second stage of degradation, when the MW of the intermediate molecules attains the some "critical" value and the products of hydrolysis become hydrophilic to provide dissolution and diffusion into water medium, the weight reduction is clearly observed at 70°C. This stage is accompanied by the changes of physical-chemical, mechanical, structural characteristics, and a geometry alteration. A similar 2-stage mechanism of PHB degradation has been described in the other publications [23, 24]. Furthermore, in the classical work of Reush [25] she showed that hydrophilization of PHB intermediates occurs at relatively low MW namely, at several decades of kDa. The results provide evidences that the reduction of MW till "critical" values to be equal about 30 kDa leads to the expansion of the second stage, namely, to the intensive weight loss.

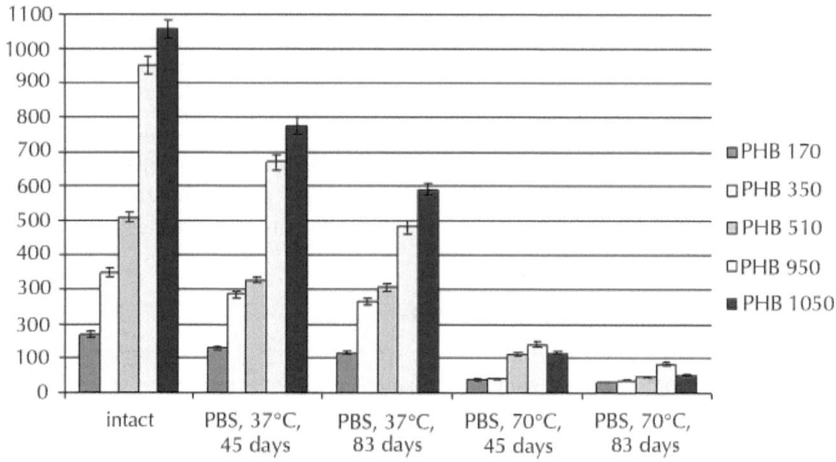

FIGURE 2 The molecular weight conversion of PHB and PHBV films during hydrolysis in phosphate buffer, pH = 7.4, 37°C and 70°C.

9.3.3 Crystallinity of PHB and PHBV

We have revealed that during hydrolytic degradation, PHB and PHBV show the MW reduction and the total weight decrease. Additionally, by the X-ray diffraction (XRD) technique we have measured the crystallinity degree of PHB and PHBV that varied depending on time in the interval of values 60–80% (see Figure 3(A)). We have noted that on the initial stage of polymer exposition to the aqueous buffer solution (at 37°C for 45 days) the crystallinity degree has slightly increased and then, under following exposition to the buffer, this characteristic is constant or even slightly decreased showing a weak maximum. When taken into account that at 37°C the total weight for the PHB films with MWs equal 350, 500, and 1000 kDa and the PHBV film with MW equals 1050 are invariable, a possible reason of the small increase in crystallinity is recrystallization described earlier for PLA [26]. Recrystallization (or additional crystallization) happens in semicrystalline polymers where the crystallite portion can increase using polymer chains in adjoining amorphous phase [22].

At higher temperature of hydrolysis, 70°C, the crystallinity increment is strongly marked and has a progressive trend. The plausible explanation of this effect includes the hydrolysis progress in amorphous area of biopolymers. It is well known that the matrices of PHB and PHBV are formed by alternative crystalline and non crystalline regions, which determine both polymer morphologies and transport of aggressive medium. Additionally, we have revealed recently by H-D exchange fourier transform infrared (FTIR) technique that the functional groups in the PHB crystallites are practically not accessible to water attack. Therefore, the hydrolytic destruction and the weight decrease are predominantly developed in the amorphous part of polymer [22, 27]. Hence, the crystalline fraction becomes larger through polymer fragment desorption from amorphous phase. This effect takes place under the strong aggressive conditions (70°C) and does not appear under the physiological conditions (37°C) when the samples have invariable weight.

Owing to the longer lateral chains in PHBV, copolymerization modifies essentially the parent characteristics of PHB such as decreasing in crystallinity, the depression of melting and glass temperatures and hence, enhancing ductility and improvement of processing characteristics [14, 28, 29]. Additionally, we have founded out that the initial crystallinity of PHB films is a monotonically increased function of initial MW (see Figure 3(B)). For samples with relatively low molecular weight it is difficult to compose the perfect crystalline entities because of a relatively high concentration of terminal groups performing as crystalline defects.

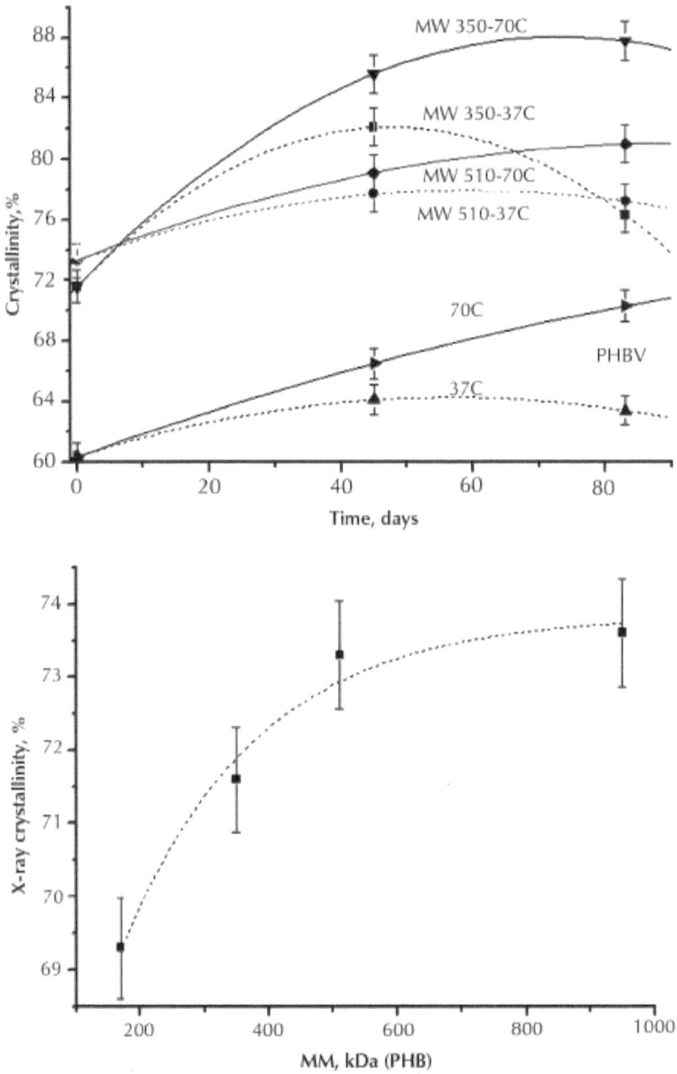

FIGURE 3 *(Continued)*

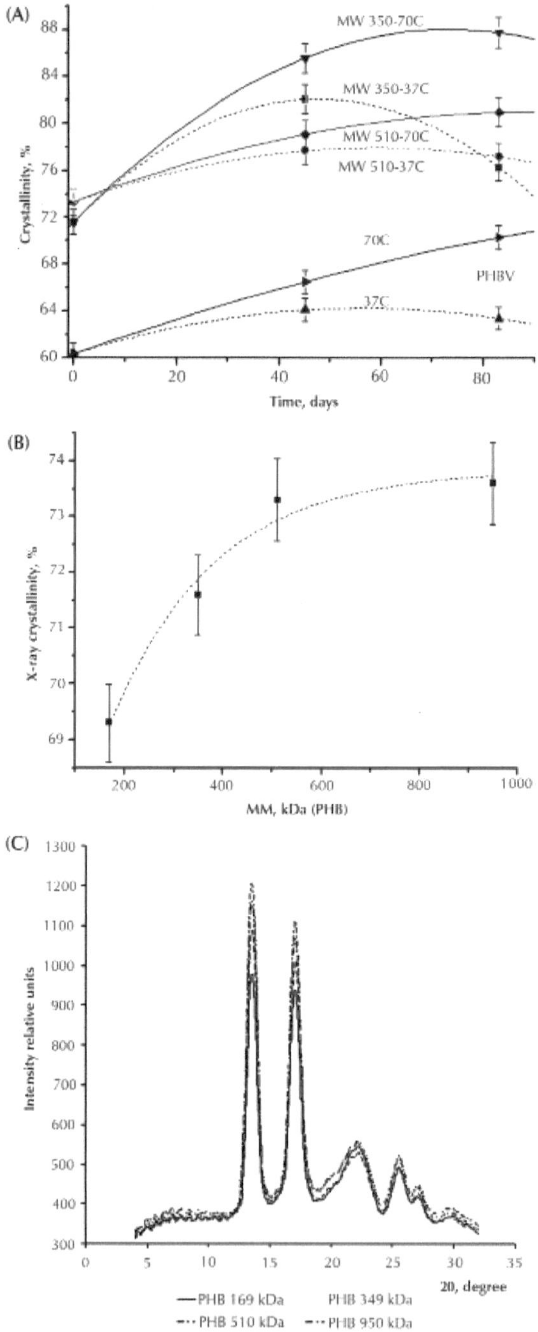

FIGURE 3 (A) Crystallinity evolution during the hydrolysis for PHB and PHBV films (denoted values of temperature and MW). (B) Crystallinity as function of initial MW for PHB films prepared by cast method. (C) X-ray diffractograms for PHB films with different MW given under x-axis.

Thus, at physiological temperature the crystallinity, measured during degradation by XRD technique has a slightly extreme character. On the initial stage of PHB degradation the crystalline/amorphous ratio is increased owing to additional crystallization through involvement of polymer molecules situated in amorphous fields. In contrast, at 70°C after reaching the critical MW values, the following desorption of water-soluble intermediates occurs. On the following stage, as the degradation is developed till film disintegration, the crystallinity drop must takes place as result of crystallite disruption.

9.3.4 The Analysis of Film Surfaces for PHB by AFM Technique

The morphology and surface roughness of PHB film exposed to corrosive medium (phosphate buffer) have been studied by the AFM technique. This experiment is important for surface characterization because the state of implant surface determines not only mechanism of degradation but the protein adsorption and cell adhesion which are responsible for polymer biocompatibility [30]. As the standard sample we have used the PHB film with relatively low MW = 170 kDa. The film casting procedure may lead to distinction in morphology between two surfaces when the one plane of the polymer film was adjacent with glass plate and the other one was exposed to air. Really, as it is shown in Figure 4 the surface exposed to air has a roughness formed by a plenty of pores with the length of 500–700 nm. The opposite side of the film contacted with glass (Figure 4(b)) is characterized by minor texture and by the pores with the less length as small as 100 nm. At higher magnification (here not presented) in certain localities it can see the stacks of polymer crystallites with width about 100 nm and length 500–800 nm.

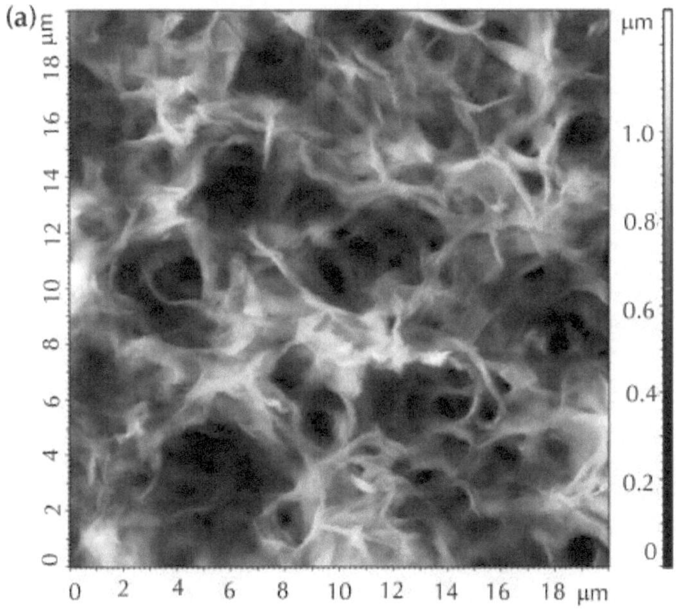

FIGURE 4 *(Continued)*

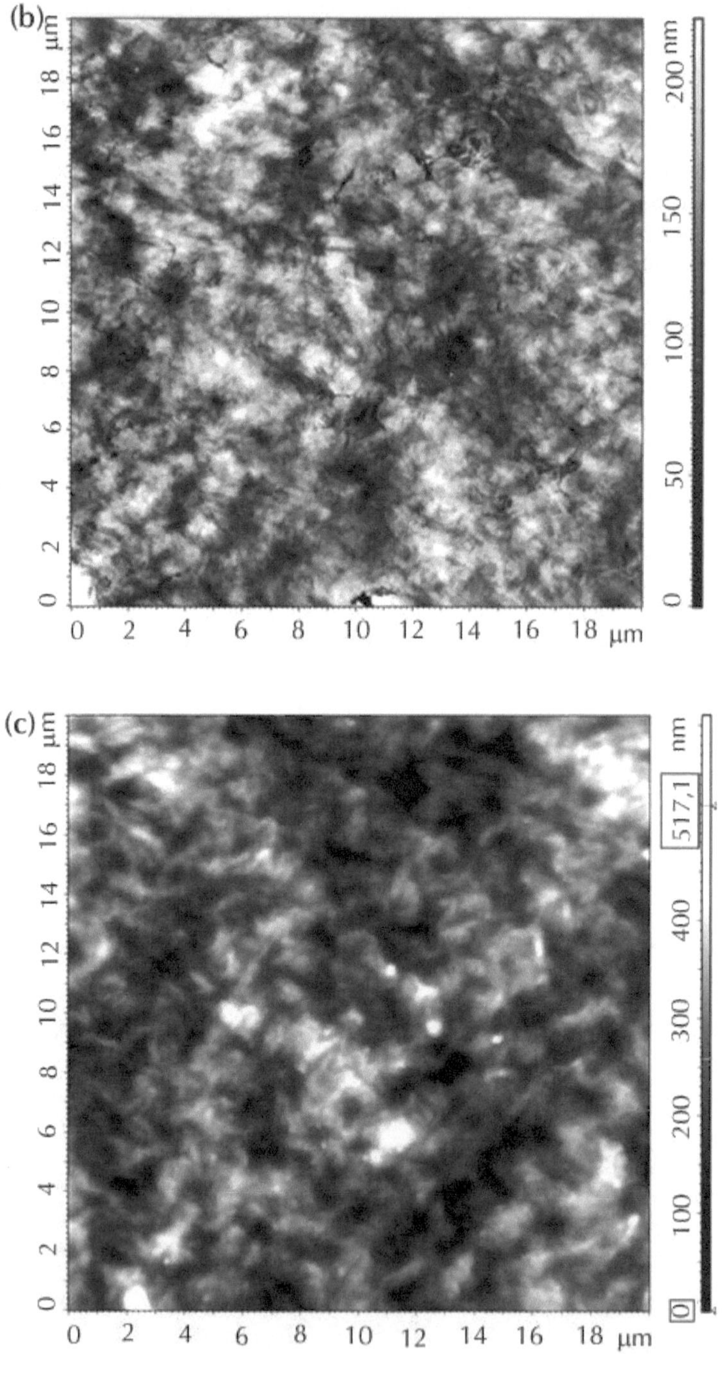

FIGURE 4 *(Continued)*

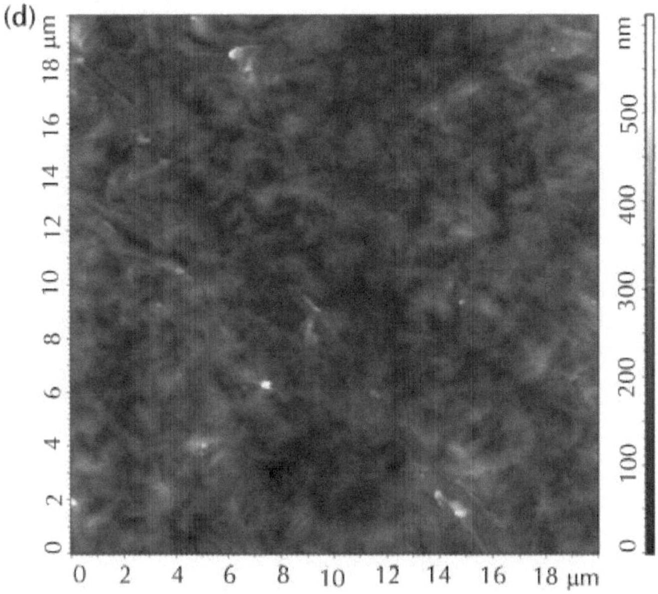

FIGURE 4 The AFM topographic images of PHB films (170 kDa) with a scan size of 18 × 18 μm: the rough surface of fresh-prepared sample (exposed to air) – (a) the smooth surface of fresh-prepared sample (exposed to glass), (b) the sample exposed to phosphate buffer at 37°C for 83 days, (c) the sample exposed to phosphate buffer at 70°C for 83 days, and (d) General magnificence is 300.

Inequality of morphology between two surfaces gets clearly evident when quantitative parameters of roughness (r_n) were compared. A roughness analysis has shown that averaged value of this characteristic:

$$R_a = \frac{1}{N}\sum_{n=1}^{N}|r_n|$$

and a root mean square roughness:

$$R_q = \sqrt{\frac{1}{N}\sum_{n=1}^{N}{r_n}^2}$$

for surfaces exposed to glass or air differ about ten times.

The variance of characteristics is related with solvent desorption conditions during its evaporation for the cast film. During chloroform evaporation from the surface faced to air, the flux forms additional channels (*viz.* the pores), which are fixed as far as the film is solidified and crystallized. Simultaneously, during evaporation the morphology and texture on the opposite side of film exposed to the glass support are not subjected to the impact of solvent transport. The morphology of the latter surface depends on energy interaction conditions (interface glass-biopolymer tension) predominantly.

The exposition of PHB film to the buffer for long time (83 days) leads to a threefold growth of roughness characteristics (see Table 1) for glass-exposed surface and practically does not affect the air-exposed surface. It is interesting that temperature of film degradation does not influence on the roughness change. The surface characteristics of film surface have the same values after treatment at 37 and 70°C (see again Table 1).

TABLE 1 Crystallinity of PHB films are exposed to phosphate buffer solution, pH = 74.

Conditions	PHB 170	PHB 350	PHB510	PHB 950	PHBV
Temperature/time					
Initial crystallinity*	69,3	71,6	73,3	73,6	60,4
37 °C 45 days	62,1	82,1	77,7	72,0	64,1
37 °C 83 days	72,6	76,3	77,2	76,2	63,4
70 °C 45 days	77,1	85,6	79,1	82,9	66,5
70 °C 83 days	-	87,8	81,0	86,1	70,3

* - standard deviations for all measurements WAXS are ± 175%.

Summarizing the AFM data we can conclude that during degradation the air-exposed, rough surface remained stable that probably related with the volume mechanism of degradation (V-mechanism [31, 32]). The pores on the surface provide the fast water diffusion into the bulk of PHB. However, under the same environmental conditions, the change of surface porosity (roughness) for glass-exposed surface is remarkable showing the engagement of surface into degradation process (S-mechanism [31, 32]). Last findings show that along with the volume processes of polymer degradation the surface hydrolysis can proceed. Several authors [20, 21] have recently reported on surface mechanism of PHB destruction but traditional point of view states a volume mechanism of degradation [12]. Here, using an advanced method of surface investigation AFM we have shown that for the same film under the same exterior conditions the mechanism of degradation could be changed depending on the prehistory of polymer preparation.

9.4 CONCLUSION

Analyzing all results related with hydrolytic degradation of PHB and its derivatives, the consecutive stages of such complicated process are presented as follows. During the initial stage, the total weight is invariable and the cleavage of biomolecules resulting in the MW decrease is observed. Within this time the PHB intermediates are too large and hydrophobic to provide solubility in aqueous media. Because the PHB crystallites stay stable, the crystallinity degree is constant as well and even it may grow up due to additional crystallization. On the second stage of hydrolysis, when the MW of intermediates attain the "critical" value, which is equal about 30 kDa, these intermediates can dissolve and diffuse from the polymer into buffer. Within this period the weight loss is clearly observed. The intensity of hydrolysis characterized by the weight loss and the MW decrement is enhanced in the series PHBV < PHB < PHB-PLA blend < PLA.

The growth of initial MW (a terminal group reducing) impacts on the hydrolysis stability probably due to the increase of crystallite perfection and crystallinity degree. The XRD data reflect this trend (see Figure 3(b)). Moreover, the surface state of PHB films explored by AFM technique depends on the condition of film preparation. After cast processing, there is a great difference in morphologies of PHB film surfaces exposed to air and to glass plate. It is well known that the mechanism of hydrolysis could include two consecutive processes: (a) volume degradation and (b) surface degradation. Under essential pore formation (in the surface layer exposed to air) the volume mechanism prevails. The smooth surface of PHB film contacted during preparation with the glass plate is degraded much intensely than the opposite rough surface (Figure 4).

We have revealed that the biopolymer MW determines the form of a hydrolysis profile (see Figure 1). For acceleration of this process we have to use the small MW values of PHB. In this case we affect both the degradation rate and the crystalline degree (Figure 3(b)). By contrast, for prolongation of service-time in a living system it is preferable to use the high-MW PHB that is the most stable polymer against hydrolytic degradation.

KEYWORDS

- **Atomic force microscopy**
- **Bacterial poly(3-hydroxybutyrate)**
- **Crystallinity**
- **Molecular weight**
- **Polylactide**
- **Wide angle X-ray scattering**

ACKNOWLEDGMENT

This work was financially supported by a special grant from the Presidium of the Russian Academy of Sciences (2011) "Academic Science to Medicine" and RAS academic project "New generation design and study of macromolecules and macromolecular structures".

REFERENCES

1. Sudesh, K., Abe, H., and Doi, Y. Synthesis, structure, and properties of polyhy-droxyalkanoates: Biological polyesters. *Progress in Polymer Science (Oxford)*, **25**(10), 1503–1555 (2000).
2. Lenz, R. W. and Marchessault, R. H. Bacterial Polyesters: Biosynthesis, Biodegradable Plastics, and Biotechnology. *Biomacromolecules*, **6**(1), 1–8 (2005).
3. Bonartsev, A. P., Iordanskii, A. L., Bonartseva, G. A., and Zaikov, G. E. Biodegradation and Medical Application of Microbial Poly (3-Hydroxybutyrate). *Polymers Research Journal*, **2**(2), 127–160 (2008).
4. Kadouri, D., Jurkevitch, E., Okon, Y., and Castro-Sowinski, S. *Critical Reviews in Microbiology*. Ecological and agricultural significance of bacterial polyhydroxyalkanoates, **31**(2), 55–67 (2005).
5. Jendrossek, D. and Handrick, R. Microbial degradation of polyhydroxyalkanoates. *Annu Rev Microbiol.*, **56**, 403–432 (2002).
6. Steinbuchel, A. and Lutke-Eversloh, T. Metabolic engineering and pathway construction for biotechnological production of relevant poly-hydroxyalkanoates in microorganisms. *Biochem. Eng. J.*, **16**, 81–96(2003).
7. Miller, N. D. and Williams, D. F. On the biodegradation of poly-beta-hydroxybutyrate (PHB) homopolymer and poly-beta-hydroxybutyrate-hydroxyvalerate copolymers. *Biomaterials*, **8**(2), 129–137 (1987).
8. Qu, X. H., Wu, Q, Zhang, K. Y., and Chen, G. Q. In vivo studies of poly(3-hydroxybutyrate-co-3-hydroxyhexanoate) based polymers: biodegradation and tissue reactions. *Biomaterials*, **27**(19), 3540–3548 (2006).
9. Fostera, L. J. R., Sanguanchaipaiwonga, V., Gabelisha, C. L., Hookc, J., and Stenzel, M. A natural-synthetic hybrid copolymer of polyhydroxyoctanoate-diethylene glycol: biosynthesis and properties. *Polymer*, **46**, 6587–6594 (2005).
10. Marois, Y., Zhang, Z., Vert, M., Deng, X., Lenz, R., and Guidoin, R. Mechanism and rate of degradation of polyhydroxyoctanoate films in aqueous media: A long-term in vitro study. *Journal of Biomedical Materials Research*, **49**(2), 216–224 (2000).
11. Freier, T., Kunze, C., Nischan, C., Kramer, S., Sternberg, K., Sass, M., Hopt, U. T., Schmitz, K. P. *In vitro* and *in vivo* degradation studies for development of a biodegradable patchbased on poly(3-hydroxybutyrate). *Biomaterials*, **23**, 2649–2657 (2002).
12. Doi, Y., Kanesawa, Y., Kawaguchi, Y., and Kunioka, M. Hydrolytic degradation of microbial poly(hydroxyalkanoates). *Makrom Chem Rapid Commun*, **10**, 227–230 (1989).
13. Renstadt, R, Karlsson, S, and Albertsson, A. C. The influence of processing conditions on the properties and the degradation of poly(3-hydroxybutyrate-co-3-hydroxyvalerate). *Macromol Symp.*, **127**, 241–249 (1998).
14. Cheng Mei-Ling, Chen Po-Ya, Lan Chin-Hung, Sun Yi-Ming. Structure, mechanical properties, and degradation behaviors of the electrospun fibrous blends of PHBHHx/PDLLA. *Polymer*, doi: 10.1016/j.polymer.2011.01.039 in press (2011).
15. Myshkina, V. L., Nikolaeva, D. A., Makhina, T. K., Bonartsev, A. P., and Bonartseva, G. A. Effect of Growth conditions on the Molecular weight of poly-3-hydroxybutyrate produced by Azotobacter chroococcum 7B. *Applied Biochemistry and Microbiology*, **44**(5), 482–486 (2008).
16. Myshkina, V. L., Ivanov, E. A., Nikolaeva, D. A., Makhina, T. K., Bonartsev, A. P., Filatova, E. V., Ruzhitsky, A. O., and Bonartseva, G. A. Biosynthesis of Poly-3-Hydroxybutyrate–3-Hydroxyvalerate Copolymer by Azotobacter chroococcum Strain 7B. *Applied Biochemistry and Microbiology*, **46**(3), 289–296 (2010).
17. Akita, S., Einaga, Y., Miyaki, Y., and Fujita, H. Solution Properties of Poly(D-β-hydroxybutyrate). 1. Biosynthesis and Characterization. *Macromolecules*. **9**, 774–780 (1976).
18. Rebrov, A. V., Dubinskii, V. A., Nekrasov, Y. P., Bonartseva, G. A., Shtamm, M., and Antipov, E. M. Structure phenomena at elastic deformation of highly oriented polyhydroxybutyrate. *Polymer Science* (Russian), **44**(A), 347–351 (2002).

19. Koyama, N. and Doi, Y. Morphology and biodegradability of a binary blend of poly((R)-3-hydroxybutyric acid) and poly((R,S)-lactic acid). *Can. J. Microbiol.*, **41**(1), 316–322 (1995).
20. Majid, M. I. A., Ismail, J., Few, L. L., and Tan, C. F. The degradation kinetics of poly(3-hydroxybutyrate) under non-aqueous and aqueous conditions. *European Polymer Journal*, **38**(4), 837–839 (2002).
21. Choi, G. G., Kim, H. W., and Rhee, Y. H. Enzymatic and non-enzymatic degradation of poly(3-hydroxybutyrate-co-3-hydroxyvalerate) copolyesters produced by Alcaligenes sp. MT-16. *The Journal of Microbiology*, **42**(4), 346–352 (2004).
22. Iordanskii, A. L., Rudakova, T. E., and Zaikov, G. E. *Interaction of polymers with corrosive and bioactive media*. VSP, New York and Tokyo (1984).
23. Wang, H. T., Palmer, H., Linhardt, R. J., Flanagan, D. R., and Schmitt, E. Degradation of poly(ester) microspheres. *Biomaterials*, **11**(9), 679–685 (1990).
24. Kurcok, P., Kowalczuk, M., Adamus, G., Jedlinrski, Z., and Lenz, R. W. Degradability of poly (b-hydroxybutyrate). Correlation with chemical microstucture. *JMS-Pure Appl. Chem*, **A32**, 875–880 (1995).
25. Reusch, R. N. Biological complexes of poly-β-hydroxybutyrate. *FEMS Microbiol. Rev.*, **103**, 119–130 (1992).
26. Molnár, K., Móczó, J., Murariu, M., Dubois, Ph., and Pukánszky, B. Factors affecting the properties of PLA/CaSO4 composites: homogeneity and interactions. *eXPRESS Polymer Letters*, **3**(1), 49–61 (2009).
27. Spyros, A., Kimmich, R., Briese, B., and Jendrossek, D. 1H NMR imaging study of enzymatic degradation in poly(3-hydroxybutyrate) and poly(3-hydroxybutyrate-co-3-hydroxyvalerate). Evidence for preferential degradation of amorphous phase by PHB depolymerase B from Pseudomonas lemoignei. *Macromolecules*, (30), 8218–8225 (1997).
28. Luizier, W. D. Materials derived from biomass/biodegradable materials. *Proc. Natl. Acad. Sci.* USA, (89), 839–842 (1992).
29. Gao, Y, Kong, L, Zhang, L, Gong, Y, Chen, G, Zhao, N, et al. *Eur Polym J*, **42**(4), 764–75 (2006).
30. Pompe, T., Keller, K., Mothes, G., Nitschke, M., Teese, M., Zimmermann, R., and Werner, C. Surface modification of poly(hydroxybutyrate) films to control cell-matrix adhesion. *Biomaterials*, **28**(1), 28–37 (2007).
31. Siepmann, J., Siepmann, F., and Florence, A. T. Local controlled drug delivery to the brain: Mathematical modeling of the underlying mass transport mechanisms. *International Journal of Pharmaceutics*, **314**(2), 101–119 (2006).
32. Zhang, T. C., Fu, Y. C., Bishop, P. L., et al. Transport and biodegradation of toxic organics in biofilms. *Journal of Hazardous Materials*, **41**(2–3), 267–285 (1995).

10 Electrochemical Methods for Estimation of Antioxidant Activity of Various Biological Objects

N. N. Sazhina, E. I. Korotkova, and V. M. Misin

CONTENTS

10.1 INTRODUCTION

For the last quarter of the century there was a considerable quantity of works devoted to research of the activity of antioxidants (AO) in herbs, foodstuff, drinks, biological liquids, and other objects. It is known that the increase in activity of free radical oxidation processes in a human organism leads to destruction of structure and properties of lipid membranes. There is a direct communication between the superfluous content of free radicals in an organism and occurrence of dangerous diseases [1, 2]. The AO are class of biologically active substances which remove excessive free radicals, decreasing the lipid oxidation. Therefore, a detailed research of the total antioxidant activity of various biological objects represents doubtless interest.

At present, there are a large number of various methods for determining the total AO content and also their activity with respect to free radicals in foodstuffs, biologically active additives, herbs, and preparations, biological liquids, and other objects [3]. However, it is impossible to compare the results obtained by different methods, since they are based on different principles of measurements, different modeling systems, and have different dimensions of the antioxidant activity index. In such cases, it is unreasonable to compare numerical values, but it is possible to establish a correlation between results obtained by different methods.

Ones from the simplest methods for study of antioxidant activity of various biological objects are electrochemical methods, in particular, ammetry and voltammetry. A comparative analysis of the AO content and their activity in juice and extracts of herbs, extracts of a tea, vegetative additives, and also in plasma of human blood is carried out in present work. Operability of methods has allowed studying also dynamics of change of the AO content and their activity in same objects during time.

Efficiency of methods has allowed studying dynamics of AO content and activity change in same objects during time. Good correlation between the total phenol antioxidant content in the studied samples and values of the kinetic criterion defining activity with respect to oxygen and its radicals is observed.

Definition of the total phenol antioxidant content and activity of compounds with respect to oxygen and its radicals in various biological objects by two electrochemical methods and analysis of obtained results.

10.2 EXPERIMENTAL PART

10.2.1 Ammetric Method for Determining the Total Content of AO

The essence of the given method consists in measurement of the electric current arising at oxidation of investigated substance on a surface of a working electrode at certain potential. An oxidation of only OH - groups of natural phenol type AO (R-OH) there is at this potential. The electrochemical oxidation proceeding under scheme $R{-}OH \rightarrow R{-}O^{\bullet} + e^{-} + H^{+}$ can be used under the assumption of authors [4], as model for measurement of free radical absorption activity which is carried out according to equation $R{-}OH \rightarrow R{-}O^{\bullet} + H^{\bullet}$. Both reactions include the rupture of the same bond O–H. In this case, the ability of same phenol type AO to capture free radicals can be measured by value of the oxidizability of these compounds on a working electrode of the ammetric detector [4].

Ammetric device "TsvetJauza-01-AA" in which this method is used to represents an electrochemical cell with a glassy carbon anode and a stainless steel cathode to which a potential 1, 3 V is applied [5]. The analyte is introduced into eluent by a special valve. As the analyte pass through the cell the electrochemical AO oxidation current is recorded and displayed on the computer monitor. The integral signal is compared to the signal received in same conditions for the comparison sample with known concentration. Quercetin and gallic acid (GA) were used in work as the comparison sample. The root-mean-square deviation (RMSD) for several identical instrument readings makes no more than 5% [5].The error in determination of the AO content including the error by reproducibility of results was within 10%. The method involves no model chemical reaction and measurement time makes 10–15 min.

10.2.2 Voltammetric Method for Determining the Total Activity of AO with Respect to Oxygen and its Radicals

The voltammetric method uses the process of oxygen electroreduction (ER O_2) as modeling reaction. This process is similar to oxygen reduction in tissues and plant extracts. It proceeds at the working mercury film electrode (MFE) in several stages with formation of the reactive oxygen species (ROS), such as $O_2^{\cdot-}$ and HO_2 [6]:

$$O_2 + e^- \rightleftarrows O_2^{\cdot-} \quad (1)$$

$$O_2^{\cdot-} + H^+ \rightleftarrows HO_2^{\cdot} \quad (2)$$

$$HO_2^{\cdot} + H^+ + e^- \rightleftarrows H_2O_2 \quad (3)$$

$$H_2O_2 + 2H^+ + 2e^- \rightleftarrows 2H_2O \quad (4)$$

For determination of total antioxidant activity it is used the ER O_2. It should be noted that AO of various natures were divided into four groups according to their mechanisms of interaction with oxygen and its radicals (Table 1) [7].

TABLE 1 Groups of biological active substances (BAS) divided according mechanisms of interaction with oxygen and its radicals.

N Group	1 group	2 group	3 group	4 group
Substance names	Catalyze, phtalocyanines of metals, humic acids.	Phenol nature substances, vitamins A, E, C, B, flavonoids.	N, S-containing substances, amines, amino acids.	Superoxide dismutase(SOD), porphyry metals, cytochrome C
Influence on ER O_2 process	Increase of ER O_2 current, potential shift in negative area	Decrease of ER O_2 current, potential shift in positive area	Decrease of ER O_2 current, potential shift in negative area	Increase in ER O_2 current, potential shift in positive area
The prospective electrode mechanism.	EC* mechanism with the following reaction of hydrogen peroxide disproportion and partial regeneration of molecular oxygen.	EC mechanism with the following chemical reaction of interaction of AO with active oxygen radicals	CEC mechanism with chemical reactions of interaction of AO with oxygen and its active radicals	EC* mechanism with catalytic oxygen reduction via formation of intermediate complex.

*The note: E – electrode stage of process, C – chemical reaction.

The first group of substances increased ER O_2 current according mechanism (5)–(7):

$$O_2 + e^- + H^+ \rightleftarrows HO_2^{\cdot} \quad (5)$$

$$HO_2^{\cdot} + e^- + H^+ \rightleftarrows H_2O_2 \quad (6)$$

$$2H_2O_2 \xrightarrow{\text{catalyst}} 2H_2O + O_2 \quad (7)$$

For the second group of the AO we suppose following mechanism of interaction of AO with the ROS (8):

$$O_2 + e^- \underset{}{\overset{k_0}{\rightleftarrows}} O_2^{\cdot -} + R\text{-}OH + H^+ \overset{k_1^*}{\rightleftarrows} H_2O_2 + R\text{–}O^{\cdot} \quad (8)$$

The third group of the BAS decreased ER O_2 current *via* the following mechanism (9)–(11):

$$O_2 \longrightarrow O_2^S \xrightarrow[k_1]{+RSH} HO_2 + \overline{e} \overset{k_{R1}}{\rightleftarrows} HO_2^{\bullet -} + RS^{\bullet} \quad (9)$$

$$HO_2^{\bullet -} + RSH \rightleftarrows H_2O_2 + RS^{\bullet} \quad (10)$$

$$RS^{\bullet} + RS^{\bullet} \rightleftarrows RS - SR \quad (11)$$

For the fourth group of substance we could suggest mechanism with catalytic oxygen reduction *via* formation of intermediate complex similar by SOD (12).

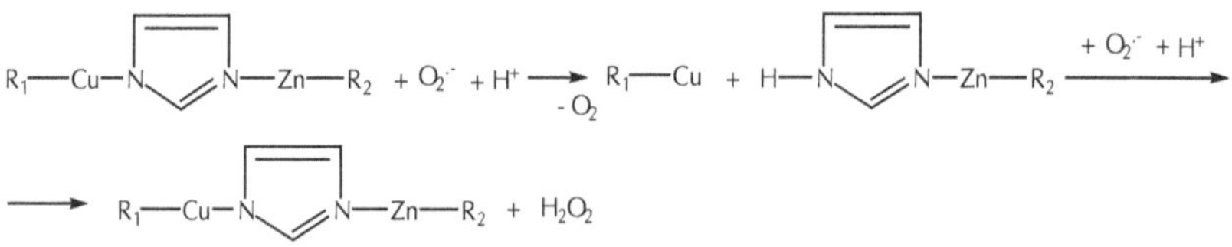

(12)

For voltammetric study of the total antioxidant activity of the samples automated voltammetric analyzer "Analyzer of TAA" (Ltd. "Polyant" Tomsk, Russia) was used. As supporting electrolyte the 10 ml of phosphate buffer (pH = 6.76) with known initial concentration of molecular oxygen was used [7]. The electrochemical cell (V = 20 ml) was connected to the analyzer and consisted of a working MFE, a silver-silver chloride reference electrode with KCl saturated ($Ag|AgCl|KCl_{sat}$) and a silver-silver chloride auxiliary electrode. The investigated samples (10–500 μl) were added in cell.

Criterion *K* is used as an antioxidant activity criterion of the investigated substances:

$$\text{For the second and third groups of AO: } K = \frac{C_0}{t}\left(1-\frac{I}{I_0}\right) \ \mu\text{mol/l}\times\text{min}, \tag{13}$$

$$\text{For the firth and four groups of AO: } K = \frac{C_0}{t}\left(1-\frac{I_0}{I}\right) \ \mu\text{mol/l}\times\text{min}, \tag{14}$$

where I, I_0—limited values of the ER O_2 current, accordingly, at presence and at absence of AO in the supporting electrolyte, C_0—initial concentration of oxygen (μmol/l), that is solubility of oxygen in supporting electrolyte under normal conditions, t — time of interaction of AOs with oxygen and its radicals, min.

This method of research has good sensitivity. It is simple and cheap. However, as in any electrochemical method of this type, the scatter of instrument readings at given measurement conditions is rather high (up to a factor of 1.5–2.0). Therefore, each sample was tested 3–5 times and results were averaged. The maximum standard deviation of kinetic criterion *K* for all investigated samples was within 30%.

10.3 DISCUSSION AND RESULTS

10.3.1 Juices of Medicinal Plant

In the present work, juices pressed out from different parts of various medicinal plants, such as basket plant or golden tendril (*Callisia fragrans*), Moses-in-the-cradle (*Rhoeo spatacea*), Dichorisandra fragrants (*Dichorisandra fragrans*), *Blossfelda kalanchoe* (*Kalanchoe blossfeldiana*), air plant (*Kalanchoe pinnatum*), and devil's backbone (*Kalanchoe daigremontiana*) were investigated [8]. To preserve the properties of the juices, they were stored in refrigerator at 12°C and unfrozen to room temperature immediately before the experiment. In both methods before being poured into the measuring cell, the test juice was diluted 100 fold. A quercetin was used as the sample of comparison. Results of measurement of the total content of AO and their activity with respect to oxygen and its radicals in juice of investigated plants received by described methods are presented in Table 2. For samples 1, 4, 7, and 12 where the small content of phenol type AO is observed, voltamperograms (VA-grams) look like characteristic for substances of the third group of Table 1 (Figure 1). Other samples show the classical phenolic mechanism as substances of the second group (Figure 2) and have high values of the AO content and kinetic criterion *K*, especially the sample 10.

TABLE 2 The total content of AOs in the juice samples and their kinetic criterion *K*.

Sample no.	Plant name	Content of AO C, mg/l	*K*, μmol/(l min)
1	Juice from *Callisia fragrans* (golden tendril) leaves	63,6	0,72
2	Juice from *Callisia fragrans* (golden tendril) lateral sprouts (4–5 mm in diameter)	279,6	1,81
3	Juice from *Rhoeo spathacea* (Moses_in_the_cradle) leaves	461,2	2,45
4	Juice from *Callisia fragrans* leaves	73,2	0,83
5	Juice from *Callisia fragrans* stalks (1–2 mm in diameter)	119,3	1,23
6	Juice from kalanchoe (*Kalanchoe blossfeldiana*) bulblets	251,3	1,78
7	Juice from *Dichorisandra fragrans* stalks (6 mm in diameter)	38,41	0,52
8	Juice from *Dichorisandra fragrans* leaves	179,1	0,77
9	Juice from *Kalanchoe pinnata* (air plant) leaves	201,5	1,99
10	Juice from *Kalanchoe daigremontiana* (devil's backbone) leaves with bulblets	742,4	4,12
11	Juice from *Kalanchoe daigremontiana* stalks (5 mm in diameter)	142,5	1,47
12	Juice from *Callisia fragrans* herb + 20% ethanol	70,2	0,88

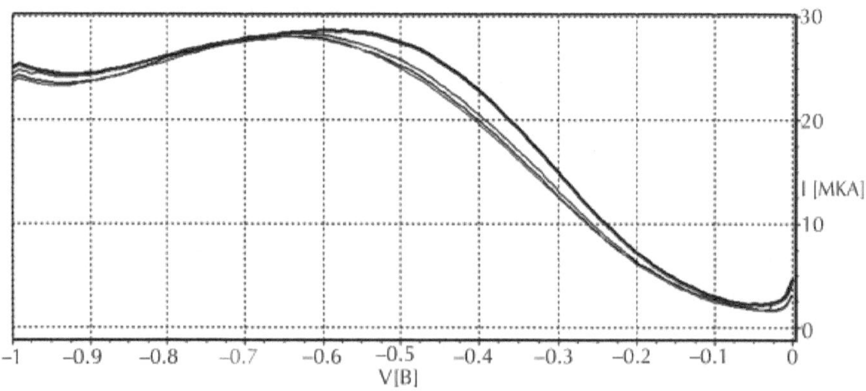

FIGURE 1 Typical VA-grams of the ER O_2 current for compounds from the third group in Table 1. The upper curve is the background current in AOs absence, left curves were recorded in AOs presence.

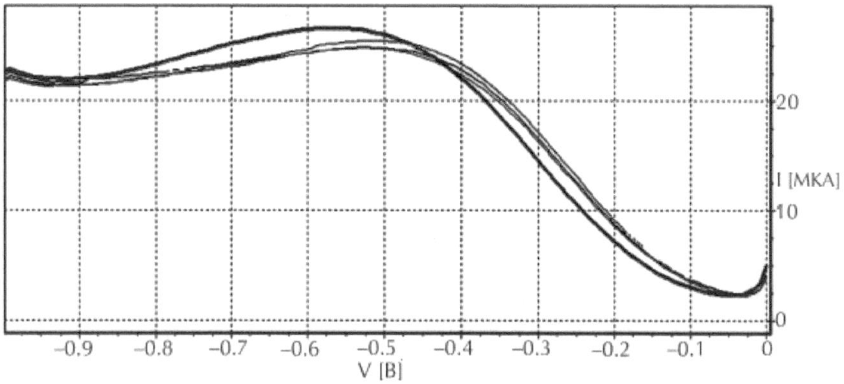

FIGURE 2 Typical VA-grams for AOs of the second substances group in Table 1. The upper curve is the background current.

Correlation dependence between the kinetic criterion *K* and the total AOs content is presented on Figure 3. The results of measurements spent for juices of medicinal plants, show high correlation (r = 0, 96) between these methods. The explanation of received results is resulted [8].

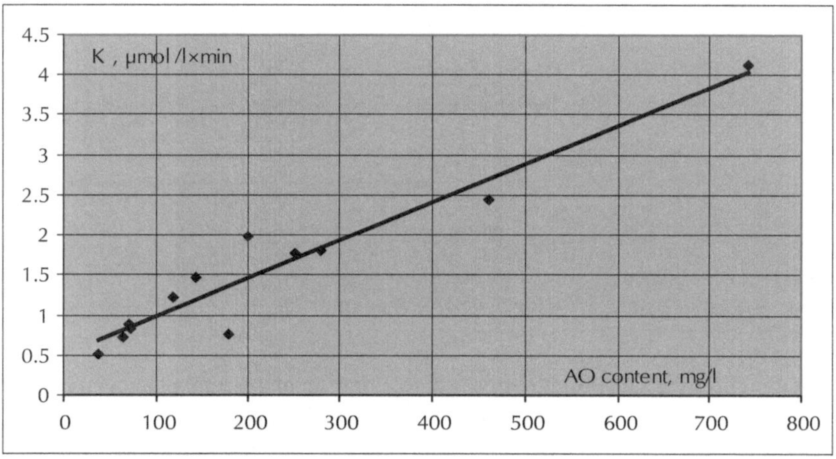

FIGURE 3 Correlation dependence between the kinetic criterion *K* and the total AOs content (r = 0, 96).

10.3.2 Water Extract of Mint

The purpose of the present work is to measure the total antioxidant activity of a water mint extract by voltammetric method and to study the mechanisms of influence of mint components on the process of ER O_2. Concurrently measurements of the total phenol AO content were carried out by an ammetric method [9]. The object of the research was water extract of the mint peppery (*Mentha piperita*). The dry herb was grinded in a mortar till particles of the size 1–2 mm. Further this herb (0.5 g) was immersed into 50 ml of distilled water with T = 95°C and held during 10 min without thermostating. Then the extract was carefully filtered through a paper filter and if necessary diluted before measurements.

The VA-grams of the ER O_2 current have been received in various times after mint extraction. It has appeared that the fresh extract (time after extraction t = 5 min) "works" on the mechanism of classical AO, reducing a current maximum and shifting its potential in positive area (Figure 2). However, approximately through t = 60 min, character of interaction of mint components with oxygen and its radicals changes, following the mechanism, characteristic for substances of the 4th group in Table 1 (Figure 4). Transition from one mechanism to another occurs approximately during t = 30 min after extraction and the kinetic criterion *K* becomes thus close to zero. At the further storage of an extract *in vitro* character of VA-grams essentially does not change and the kinetic criterion caused by other mechanism, grows to values 2.0 μmol/l·min during 3 hr after extraction.

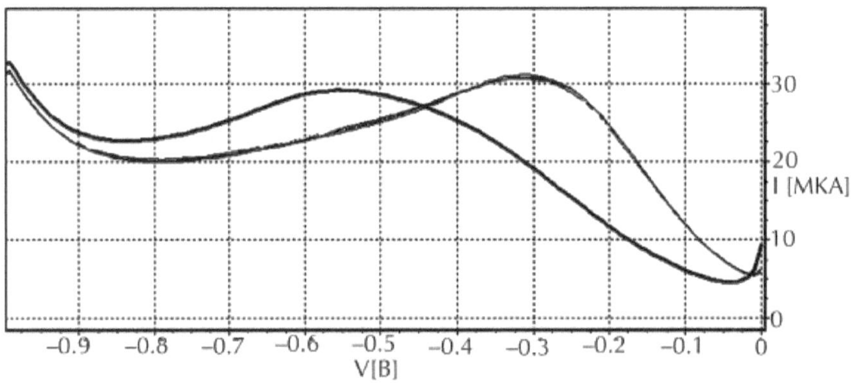

FIGURE 4 The VA-grams of the ER O_2 current in the absence (left curve) and in the presence (right curves) of mint extract at t = 60 min after mint extraction.

In parallel for the same mint extract the registration of the total phenol AO content C in mg of GA per 1 g of dry mint has been made during extract storage time t by ammetric method. Dependence C on t testifies to notable falling of C after extraction of a mint extract (approximately 20% for 2 hr of storage). It is possibly explained by destruction of unstable phenol substances contained in an extract. Therefore, apparently, VA-grams character and *K* values during the first moment after extraction could be established as influence of classical phenol AOs on the ER O_2 process. The general character of ER O_2 is defined already by the mint substances entering into 4th group of the Table 1. Probably, various metal complexes as mint components are dominated causing increase of ER O_2 current and catalytic mechanism of ER O_2. More detailed statement of experimental materials given [9].

10.3.3 Extracts of Tea, Vegetative Additives, and their Mixes

In the given results of measurements of the total AO content and activity by two methods in water extracts of some kinds of tea and vegetative additives are presented [10]. These parameters were measured also in extracts of their binary mixes to study possible interference of mixes components into each other. Objects of research were water extracts of three kinds of tea (Chinese green tea "Eyelashes of the Beauty", gray tea with bergamot "Earl gray tea" and black Ceylon tea "Real"), mint peppery (*Mentha piperita*) and the dry lemon crusts. The ten extracts of binary mixes of the listed samples with a different weight parity of components have been investigated also. Preparation of samples and extraction spent in the same conditions as for mint. The GA was used as the comparison sample in an ammetric method. Efficiency of this method has allowed tracking the dynamics of AO content change in investigated samples directly after extraction. On Figure 5 dynamics of total phenol AO content change for five samples is shown. The most considerable content C decrease is observed in tea extracts (20–25%) that possibly as well as for mint is explained by destruction of the unstable phenol substances in extracts (katehins, teaflavins, tearubigins, etc.). For extract of lemon crusts the total AO content is much less and practically does not change during first minutes after extraction.

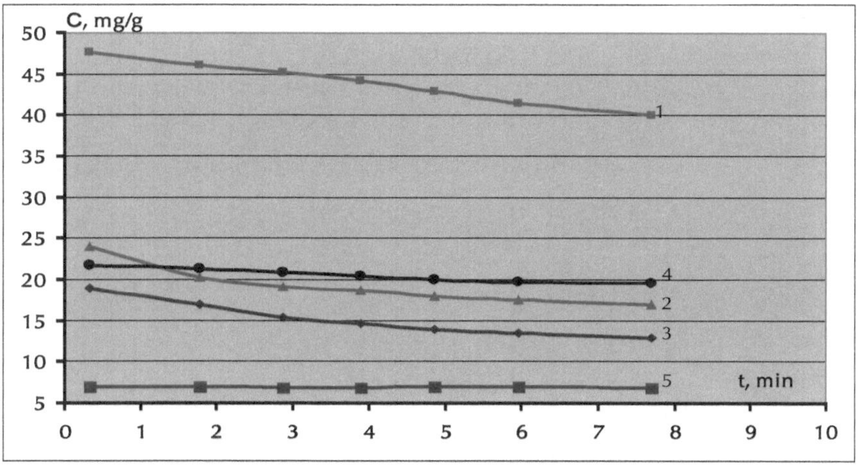

FIGURE 5 Dynamics of the total AO content *C* change (in units of gallic acid) during extract storage time t: 1–green Chinese tea, 2–gray tea with bergamot, 3–black Ceylon tea, 4–mint, and 5–lemon crusts (for lemon crusts C was increased in 5 times).

Measurement results of the total AO content for extracts of tea and additives (Figure 6(a)) and extracts of their mixes in a different parity (Figure. 6(b)) are presented. The AO content in mixes (c) is calculated under the additive AO content contribution of mix components taken from Figure 6(a) according to their parity.

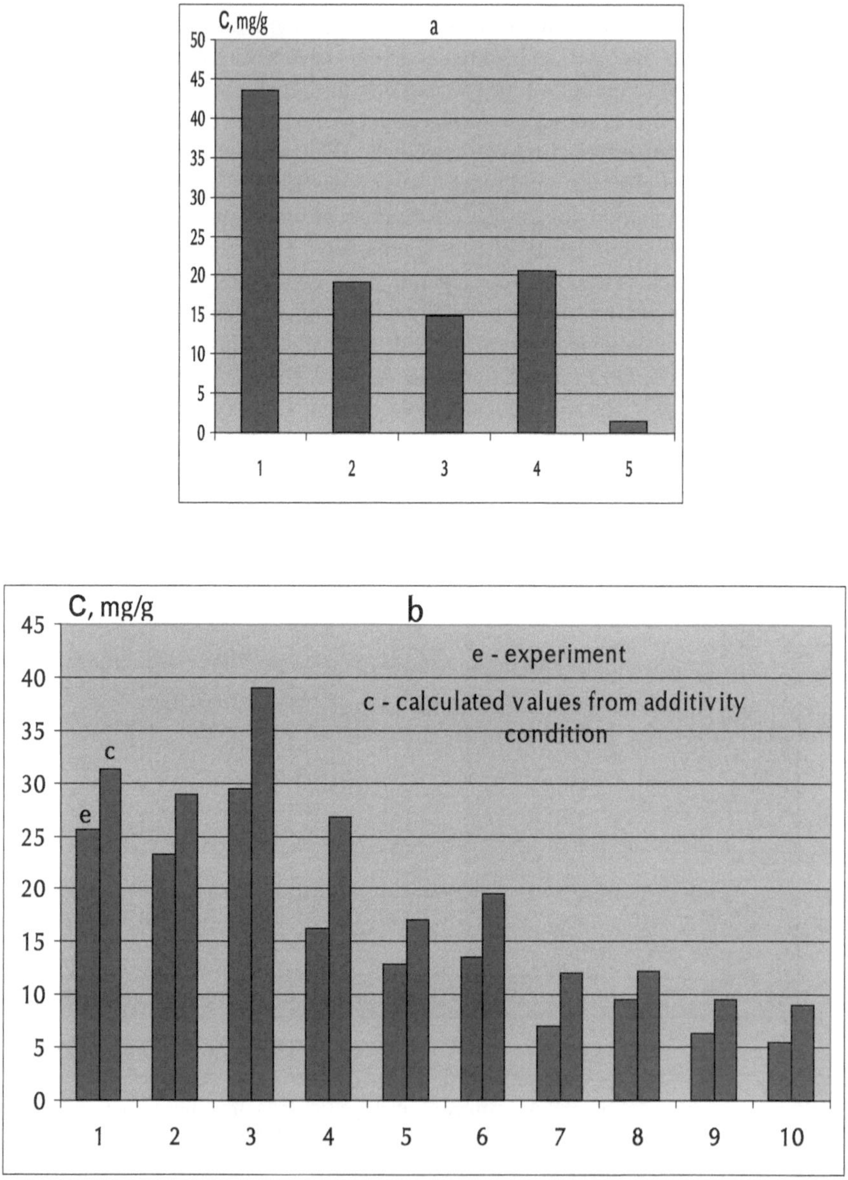

FIGURE 6 The total AO content *C* (in units of gallic acid): a - in extracts of: 1–green Chinese tea, 2–gray tea with bergamot, 3–black Ceylon tea, 4–mint, and 5–lemon crusts. b - in extracts of tea and additives mixes: 1–tea 1 + tea 2 (1:1), 2–tea 1 + tea 3 (1:1), 3–tea 1 + mint (4:1), 4–tea 1 + lemon crusts (3:2), 5–tea 2 + tea 3 (1:1), 6–tea 2 + mint (4:1), 7–tea 2 + lemon crusts (3:2), 8–tea 3 + mint (4:1), 9–tea 3 + lemon crusts (3:2), and 10–mint (4:1) + lemon crusts (2:3). In brackets the parity between components of mixes is specified.

As to mixes of tea and additives (Figure 6(b)) the measured values of the AO content in extracts of investigated mixes (e) have considerable reduction in comparison with the additive contribution of the phenol AO content of mix components that is observed their strong antagonism. Especially, it is considerable for extracts of tea with lemon crusts mixes.

The measured values of the total AO activity with respect to oxygen and its radicals *K* are presented for extracts of tea, additives (Figure 7(a)) and their mixes (Figure 7(b)).

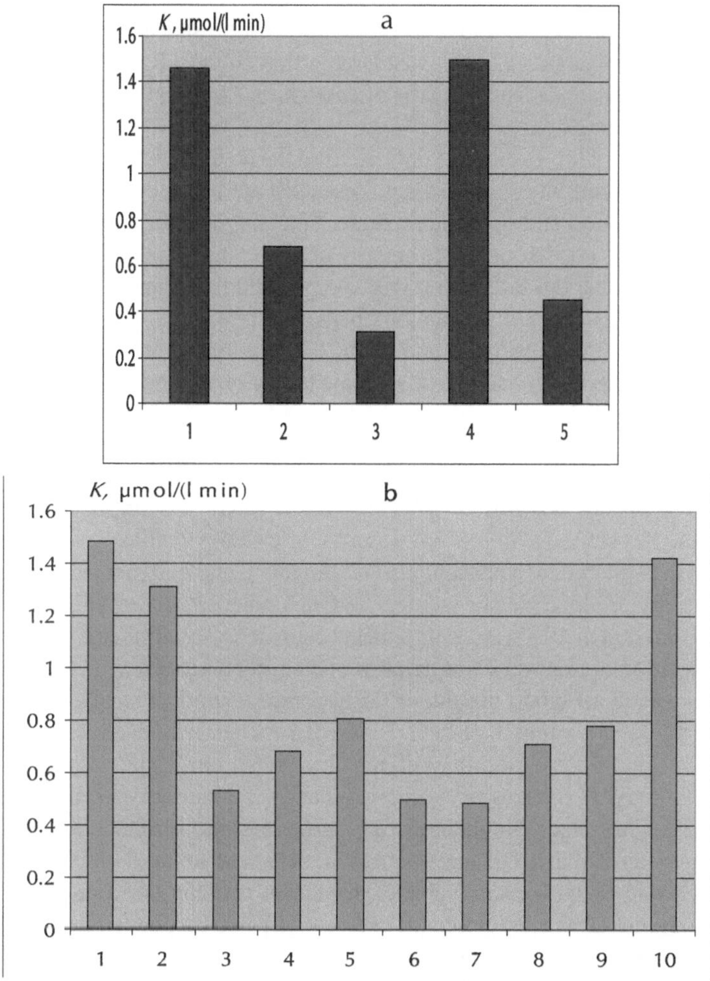

FIGURE 7 The total AO activity with respect to oxygen and its radicals *K*:*a* - in extracts of tea and additives: 1–green Chinese tea, 2–gray tea with bergamot, 3–black Ceylon tea, 4–mint, and 5 – lemon crusts. *b* - in extracts of tea and additives mixes: 1–tea 1 + tea 2 (1:1), 2–tea 1 + tea 3 (1:1), 3–tea 1 + mint (4:1), 4–tea 1 + lemon crusts (3:2), 5–tea 2 + tea 3 (1:1), 6–tea 2 + mint (4:1), 7–tea 2 + lemon crusts (3:2), 8–tea 3 + mint (4:1), 9–tea 3 + lemon crusts (3:2), and 10–mint (4:1) + lemon crusts (2:3). In brackets the parity between components of mixes is specified.

For all samples of extracts, except an extract of the lemon crusts, dominating character of interaction of extract components with oxygen and its radicals has not phenolic, but the catalytic nature. This interaction proceeds on the mechanism (12), characteristic for substances of 4th group in Table 1. Lemon crusts extract "works" on the mechanism of classical AO (8). Unlike values of the phenolic AO content measured by a ammetric method. The kinetic criterion has appeared maximum not only for extract of green tea but also for mint extract, minimum – for extract of black tea. In spite of the fact that the phenol AO content in extracts of tea and mint has appeared more than in lemon crusts, final total activity of tea and mint extracts is defined not by phenol type substances, but, apparently, various metal complexes, present in them, and catalyze of proceeding chemical processes. Considerable shift of the ER O_2 current maximum potential in positive area attests in favor of enough high content of phenol substances in teas and mint extracts.

For the activity of mixes extracts deviations of the measured values *K* from the values calculated on additivity (here are not presented) are big enough and are observed both towards reduction and towards increase. For this method, apparently, the additively principle does not "work" since activity of mix components has the different nature and the mechanism of interaction with oxygen and its radicals. Activity of mixes is defined not only chemical interactions between substances, but diffusion factors of these substances to an electrode and so on. Activity of tea mixes extracts changes weaker in comparison with activity of separate tea extracts. The possible explanation of the received results is presented [10].

10.3.4 Plasma of Human Blood

A human blood plasma is a difficult substance for researches. Its antioxidant activity is defined, mainly by presence in it of amino acids, uric acid, vitamins E, C, glucose, hormones, enzymes, inorganic salts, and also intermediate and end metabolism products. The total activity of blood plasma is integrated parameter characterizing potential possibility of AO action of all plasma components considering their interactions with each other. The purpose of this work was measurement of total activity with respect to oxygen and its radicals of blood plasma of thirty persons simultaneously with measurement of the total AO content in plasma. Blood plasma has been received by centrifuging at 1,500 r/min of blood of thirty patients from usual polyclinic with different age, sex, and pathology. It is necessary to notice that for the majority of plasma samples VA-grams were stable for 3-4 identical measurements and resulted on 2nd group of Table 1 (Figure 2). The ER O_2 current potential shift was small (0–0, 03 V) that testifies to presence in plasma of small phenol substances content. For some blood plasma samples (4 samples from 30) VA-gram character was corresponded to substances of 4th group in Table 1. For studying of correlation of the results received by two methods, values of *K*, measured during 30 min after plasma defrosting have been selected for the samples having VA-grams of 2nd group. On Figure 8 these values are presented together with corresponding measured values of the total phenol AO content C, spent also during 30 min after plasma defrosting. Correlation of received results with factor r = 0, 81 is observed. It means that in blood plasma of many patients there are phenol substances which define dominant processes of plasma components interaction with oxygen and its radicals. Results of this work were published [11].

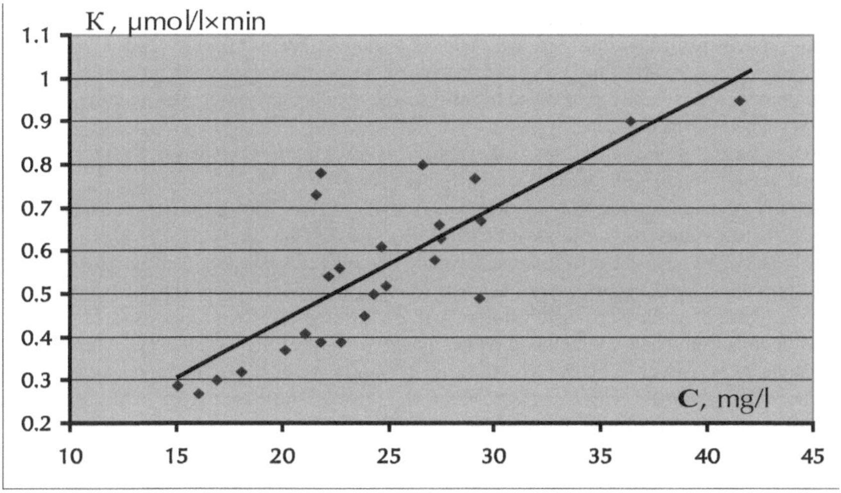

FIGURE 8 Correlation dependence between the kinetic criterion *K* and the total AO content C (in units of gallic acid) for samples of blood plasma having VA-grams of 2nd type. r = 0, 81.

10.4 CONCLUSION

Use of two operative electrochemical methods realized in devices "TsvetJauza-01-AA" and "Analyzer of TAA" allows quickly and cheaply to define the total content of AO and their activity with respect to oxygen and its radicals in various biological objects. Results of the present work show good correlation of these methods and the specified devices can be applied widely in various areas.

KEYWORDS

- **Additive**
- **Ammetry**
- **Antioxidants**
- **Blood plasma**
- **Herbs**
- **Tea**
- **Voltammetry**

REFERENCES

1. Study of Synthetic and Natural Antioxidant *in vivo* and *in vitro*. E. B. Burlakova (Ed.). Nauka, Moscow (1992).
2. Vladimirov, Yu. A. and Archakov, A. I. *Lipid peroxidation in biologicall membranes*. Nauka, Moscow (1972).

3. Roginsky, V. and Lissy, E. Review of methods of food antioxidant activity determination. *Food Chemistry*, **92**, 235–254 (2005).
4. Peyrat_Maillard, M. N., Bonnely, S., and Berset, C. Determination of the antioxidant activity of phenolic compounds by coulometric detection. *Talanta*, **51**, 709–714 (2000).
5. Yashin, A. Ya. Inject-flowing system with ammetric detector for selective definition of antioxidants in foodstuff and drinks. *Russian chemical magazine*, **52**(2), 130–135 (2008).
6. Korotkova, E. I., Karbainov, Y. A., and Avramchik, O. A. Investigation of antioxidant and catalytic properties of some biological-active substances by voltammetry. *Anal. and Bioanal. Chem.*, **375**(1–3), 465–468 (2003).
7. Korotkova, E. I. *Voltammetric method of determining the total AO activity in the objects of artificial and natural origin*. Doctoral thesis, Tomsk (2009).
8. Misin, V. M. and Sazhina, N. N. Content and Activity of Low_Molecular Antioxidants in Juices of Medicinal Plants. *Khimicheskaya Fizika*, **29**(9), 1–5 (2010).
9. Sazhina, N., Misin, V., and Korotkova, E. Study of mint extracts antioxidant activity by electrochemical methods. *Chemistry and Chemical Technology*, **5**(1), 13–18 (2011).
10. Misin, V. M., Sazhina, N. N., and Korotkova, E. I. Measurement of tea mixes extracts antioxidant activity by electrochemical methods. *Khim. Rastit. Syr'ya*, (2), 137–143 (2011).
11. Sazhina, N. N., Misin, V. M, and Korotkova, E. I. *The comparative analysis of the total content of antioxidants and their activity in the human blood plasma*. Theses of reports of 8th International conference "Bioantioxidant", Moscow, pp. 301–303 (2010).

11 Development of Conducting Polymers: (Part I)

A. K. Haghi and G. E. Zaikov

CONTENTS

11.1 INTRODUCTION

Electrically conducting polymers still are interest of many researchers due to their various applications such as rechargeable batteries, antistatic coating, shielding of electromagnetic interferences, sensors, and so on [1]. Among them, polypyrrole (PPy) has attracted great attention since it exhibits a high electrical conductivity, electrochemical activity, and good environmental stability. The PPy is polymerized by either conventional electrochemical or chemical polymerization methods in insoluble or infusible due to the strong inter or intramolecular interactions. The poor processability, brittleness, insolubility, and unstable electrical properties of PPy have limited its practical applications.

In this chapter, Lycra/Polyester fabric has been investigated—Pyrrole monomer purchased from sigma Aldrich chemical company, was distilled before use and was stored in a freezer—Naphtalene disulfonic acid (NDSA), ferric chloride were produced from Merck as an dopant and oxidant—Deionized water sulfuric acid from Merck, have been used.

11.2 EXPERIMENTAL

The samples of Lycra/Polyester fabrics were first pretreated in 1M solution of sulfuric acid for 30 min at room temperature.

All samples were then chemically polymerized in an aqueous solution containing 0.015M Pyrrole, 0.04M Ferric chloride and 0.005M NDSA, respectively as monomer, oxidant and dopant at room temperature but at various time of polymerization reaction. Different reaction times varied from 1 to 6 hr were investigated:

(1) 1 hr
(2) 2 hr
(3) 3 hr
(4) 4 hr
(5) 6 hr.

The PPy deposited on the fabrics surface then the black conductive fabrics were washed with deionized water and dried in desiccators at room temperature (in order to avoid oxidative reaction in the air), and in oven at 40°C.

11.2.1 Instrumentation

The electrical conductivity of prepared fabrics was measured by four probe technique, (According to ASTM standard F43-93). Schematic representation of this technique is shown in Figure 1.

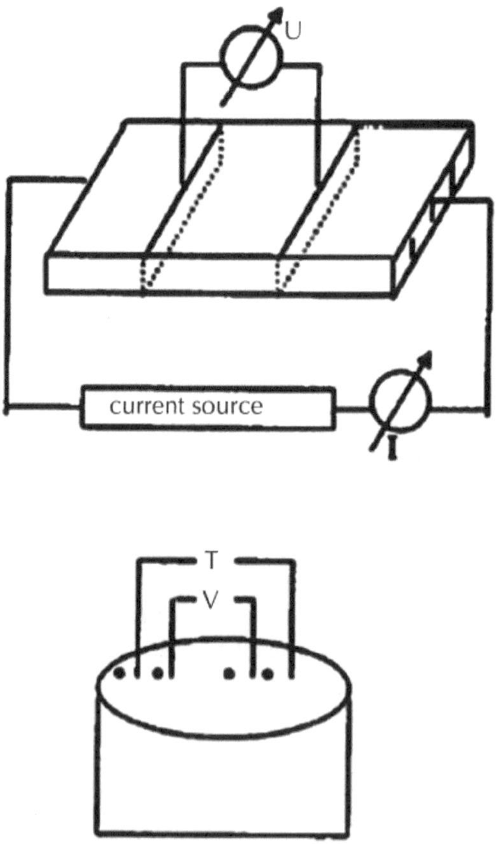

FIGURE 1 Resistance Measurement using four probe techniques.

The contacts and that of the wires leading to the voltmeter, it is necessary to keep the current passing through these contacts and wires sufficiently low. In this way, the current to the sample is supplied by two separate contacts. The conductivity of the sample is calculated by the formula as given:

$$\sigma = \frac{1}{2\pi s} \cdot \frac{i}{V}$$

s = Spacing between probes (usually 0.1 cm).
i = Current passed through outer probes.
V = Voltage drop across inner probes.

11.3 DISCUSSION AND RESULTS

Chemical polymerization is carried out by mixing monomer, dopant, and oxidant in a suitable solvent. The chemical method enables the preparation of these polymers on a large scale without requiring special equipment. The polymerization of pyrrole proceeds through the formation of radical cations that couple to form oligomers which are further oxidized to form additional radical cations.

The reaction time is an important factor to determine the final properties of the conductive textile obtained with chemical polymerization process. So in this work, the effect of reaction time using to prepare PPy coated Lycra/Polyester fabric pretreated using *in situ* chemical oxidation process has been investigated.

The electrical conductivity of all coated samples are showed and compared in Figure 2.

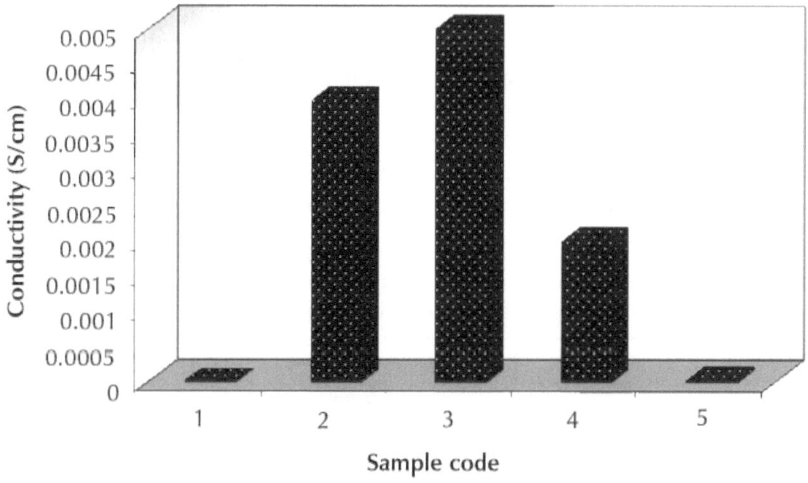

FIGURE 2 Electrical conductivity of Lycra/Polyester PPy coated fabrics pretreated, prepared with different polymerization reaction times.

Results show that electrical conductivity is dependent to reaction time. The electrical conductivity of the PPy coated Lycra/Polyester fabrics varies from 5×10^{-4} S/cm (after 1 hr) to 5×10^{-3} S/cm (after 3 hr) and then decrease to 1.5×10^{-5} S/cm (after 6 hr).

This means that polymerization reaction is nearly completed at about 3 hr. So, we suggest that the reaction time can be choose to be 3 hr because of the highest electrical conductivity will be obtained.

It founds that the thickness of the PPy coating increased with reaction time to 3 hr as depicted by the change of the color from gray, brown to dark, and at this time, no polymer was found in the solution. But application of longer time in process causes products take place in an dramatist environment (for example to study Launder fastness) with uncontrollable conditions, such as time, material, and so on. In the other hand, taking place of products in undesirable conditions after completing polymerization caused negative effect on electrical conductivity of products.

Furthermore, because of accomplish the polymerization reaction in acid solution (PH = 5), final products must be come out from reaction solution immediately to avoid unwished reactions (Figure 3), this action caused conductivity of products remains stable.

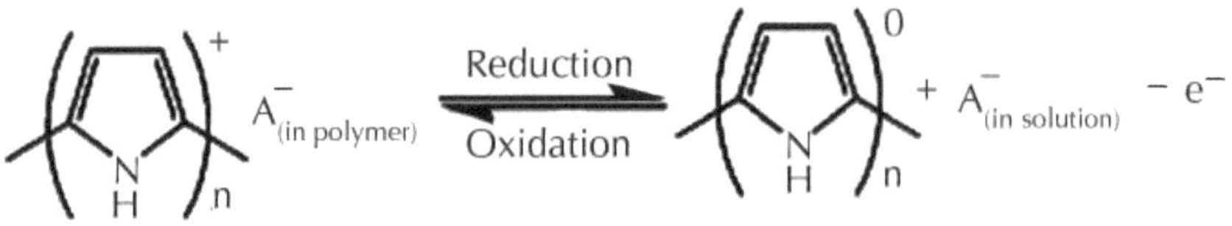

FIGURE 3 Oxidation process of pyrrole monomer.

Really, H^+ ions existing in polymerization solution (even small quantity) are obtrusive to conductive products and led the original reaction to reverse direction that decreases the electrical conductivity of final products.

In following table conductivity of PPy coated fabrics pretreated, have compared from various desiccate temperature. According that Table 1, we can say that difference of desiccate temperature has no sensible effect to final conclude.

TABLE 1 Comparison the electrical conductivity of PPy coated fabrics pretreated, prepared with various polymerization reaction times, for purposes of desiccate temperature.

Reaction time:	Desiccator (room temp)	oven (40°C)
1 hr	5.0×10^{-4}	1.0×10^{-4}
2 hr	4.0×10^{-3}	3.0×10^{-3}

TABLE 1 *(Continued)*

Reaction time:	Desiccator (room temp)	oven (40°C)
3 hr	5.0×10^{-3}	3.7×10^{-3}
4 hr	2.0×10^{-3}	1.0×10^{-3}
6 hr	1.5×10^{-3}	1.0×10^{-3}

11.4 CONCLUSION

The research on conducting polymers began nearly a quarter century ago when films of acetylene were found to exhibit dramatic increase in electrical conductivity when exposed to iodine vapors. Many other small conjugated molecules were found to polymerize conducting conjugated polymers which were either insulating or semi conductive but becoming conductive upon oxidation or reduction. Among the conducting polymers, the PPy is widely used to prepare conducting textiles due to the unique ability of the polymerizing species to absorb onto a hydrophobic surface and form uniform coherent films using *in situ* chemical polymerization.

In the present work, an attempt has been made to study on the effect of reaction time in chemical oxidation polymerization of pyrrole monomer on the electric performance of PPy coated Lycra/Polyester fabric pretreated.

Results show that when the *in situ* polymerization of pyrrole monomer was happened at 3 hr, the electrical conductivity of PPy coated fabric pretreated increased more than other conditions, so this reaction time can be choose as optimum.

The Lycra/Polyester fabric coated PPy so prepared exhibit a suitable value of conductivity equal to 5×10^{-3} S/cm and are expected to find vast potential of application.

KEYWORDS

- **Chemical oxidation**
- **Four probe technique**
- **Naphtalene disulfonic acid**
- **Polyester fabric**
- **Polypyrrole**

REFERENCES

1. Tourillon, G. Handbook of conducting polymers. T. A. Skotheim (Ed.). *Marcel Dekker*, New York, **1**, 293 (1986).
2. Scrosati, B. *Science and applications of conducting polymers*. Chapman and Hall, London ch. 7 (1993).
3. Jasne, S. *Encyclopedia of polymer science and engineering*. John Wiley, New York p. 51 (1988).

4. Om Bockris, J. and Miller, D. *Conducting polymers Special applications*. L. Alcacer (Ed.). Reidel, Dordrecht p. 1 (1989).
5. Palmisano, F., De Benedetto, G. E., and Zambonin, C. G. *Analyst*, **122**, 365 (1997).
6. Kumar, D. and Sharma, R. C. *Eur. Polym. J.*, **34**(8), 1053 (1998).
7. Chen, S. A. and Tsai, C. C. *Macromolecules*, **26**, 2234 (1991).
8. Wei, Y., Tian, J., MacDiarmid, A. G., Masters, J. G., Smith, A. L., and Li, D. *J. Chem. Soc. Chem. Commun.*, **7**, 552 (1994).
9. Beadle, P., Armes, S. P., Gottesfeld, S., Mombourquette, C., Houlton, R., Andrew, W. D., and Agnew, S. F. *Macromolecules*, **25**, 2526 (1992).
10. Somnathan, N. and Wehner, G. *Ind. J. Chem.*, **33**(A), 572 (1994).
11. Skotheim, T. A. In Handbook of Conducting Polymers. *Marcel Dekker*, New York, **1–2**, (1986).
12. Malhotra, B. D., Ghosh, S., and Chandra, R. *J. Appl. Polym. Sci.*, **40**, 1049 (1990).
13. Neglur, B. R., Laxmeshwar, N. B., and Santanan, K. S. V. *Ind. J. Chem.*, **33**(A), 547 (1994).
14. Curran, D., Grimshaw, J., and Perera, S. D. *Chem. Soc. Rev.*, **20**, 1 (1991).
15. Malli, S. *Ind. J. Chem.*, **33**(A), 524 (1994).
16. Annapoorni, S., Sundaresan, N. S., Pandey, S. S., and Malhotra, B. D. *J. Appl. Phys.*, **74**, 2109 (1993).
17. Jolly, R., Petrescu, C., Thiebelmont, J. C., Marechal, J. C., and Menneteau, F. D. *Journal of coated fabrics*, pp. 228–236 (January 23, 1994).
18. Jin, X. and Gong, K. *Journal of coated fabrics*, **26**, 36–43 (1996).
19. Trivedi, D. C. and Dhawan, S. K. In proc. of polymer symp. S. Sivaram (Ed.). McGrow Hill, Pune, India (1991).
20. Gregory, R. V., Kimbrell, W. C., and Kuhn, H. H. *Journal of coated fabrics*, **20**, (January, 1991).

12 Development of Conducting Polymers: (Part II)

A. K. Haghi and G. E. Zaikov

CONTENTS

12.1 INTRODUCTION

Electrical conductivity of conductive polymers may increase when the polymers tend to nanoscale [1-9]. Systems in nanoscale may possess entirely new physical and chemical characteristics for example, higher electrical conductivity arises when the size of a wire is reduced below certain critical thickness (nanoscale). Using these properties raises the potential of electrospinning to make fibers at the nano level with unusual properties that is impossible at the level of the visible world [9-19].

In the present work, we used polyaniline (PANI)/polyacrylonitrile (PAN) blend to form a nonwoven mat. The PANI exists in a large number of intrinsic redox states. The half oxidized emeraldine base is the most stable and widely investigated state in the PANI family that can be dissolved in N-methyl-2 pyrolidon (NMP). The polyaniline emeraldine base (PANIEB)/polyacrilonitrile blend solution in NMP was prepared and then it was electrospun with different blending ratio. The fibers diameter, fibers morphology, and electrical conductivity of the mats were analyzed and discussed.

12.2 EXPERIMENTAL

The commercial PAN polymer containing 6% mathtlacrylate with molecular weight (Mw) of 100,000 was supplied by Polyacryl Iran Co (IRAN). The NMP was from Riedel-de Haën. Aniline from Merck was vacuum distilled prior to use. The PANI used was synthesized in laboratory.

The PANI was synthesized by the oxidative polymerization of aniline in acidic media. The 3 ml of distilled aniline was dissolved in 150 ml of 1N HCl and kept at 0–5°C. 7.325g of $(NH_4)_2S_2O_8$ was dissolved in 35 ml of 1N HCl and added drop wise under constant stirring to the aniline/HCl solution over a period of 20 min. The resulting dark green solution was maintained under constant stirring for 4 hr. The prepared suspension was dialyzed in a cellulose tubular membrane (Dialysis Tubing D9527, molecular cutoff= 12,400, Sigma) against distilled water for 48 hr. Then it was filtered and washed with water and methanol. The synthesized PANI was added to 150 ml of 1N (NH_4) OH solution. After an additional 4 hr the solution was filtered and a deep blue emeraldine base form of PANI was obtained PANIEB. The synthesized PANI was dried and crushed into fine powder and then passed through a 100 mesh. Intrinsic viscosity of the synthesized PANI dissolved at sulfuric acid (98%) was 1.18 dl/g at 25°C.

The PANI solution with concentration of 5% (W/W) was prepared by dissolving exact amount of PANI in NMP. The PANI was slowly added to the NMP with constant stirring at room temperature. This solution was then allowed to stir for 1 hr in a sealed container. The 20% (W/W) solution of PAN in NMP was prepared separately and was added drop wise to the well-stirred PANI solution. The blend solution was allowed to stir with a mechanical stirrer for an additional 1 hr.

Various polymer blends with PANI content ranging from 10 to 30 wt% were prepared by mixing different amount of 5% PANI solution and 20% PAN solution. Total concentration of the blend solutions were kept as 12.5%.

12.3 ELECTROSPINNING

Polymeric nanofibers can be made using the electrospinning process which has been described [20-35]. Electrospinning, Figure1 uses a high electric field to draw a polymer solution from tip of a capillary toward a collector. A voltage is applied to the polymer solution which causes a jet of the solution to be drawn toward a grounded collector. The fine jets dry to form polymeric fibers which can be collected as a web.

The electrospinning equipment used a variable high voltage power supply from Gamma High Voltage Research (USA). The applied voltage can be varied from 1 to 30 kV. A 5 ml syringe was used and positive potential was applied to the polymer blend solution by attaching the electrode directly to the outside of the hypodermic needle with internal diameter of 0.3 mm. The collector screen was a 20 × 20 cm alluminum foil which was placed 10 cm horizontally from the tip of the needle. The electrode of opposite polarity was attached to the collector. A metering syringe pump from New Era pump systems Inc. (USA) was used. It was responsible for supplying polymer solution with a constant rate of 20 μl/min.

Electrospinning was done in a temperature controlled chamber and temperature of electrospinning environment was adjusted on 25, 50, and 75°C. Schematic diagram of the electrospinning apparatus was shown in Figure 1. Factorial experiment was designed to investigate and identify the effects of parameters on fiber diameter and morphology (Table 1).

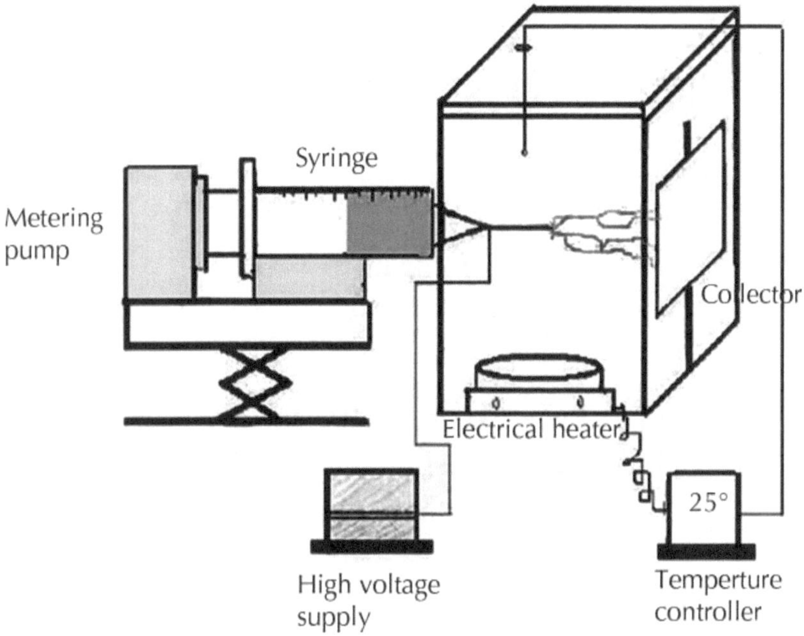

FIGURE 1 Schematic diagram of electrospinning apparatus.

12.4 CHARACTERIZATION

Shear viscosities of the fluids were measured at shear rate of 500 $s^{-1}$and 22°C using a Brookfield viscometer (DVII+, USA). Fiber formation and morphology of the electrospun PANI/PAN fibers were determined using a scanning electron microscope (SEM) Philips XL-30A (Holland). Small section of the prepared samples was placed on SEM sample holder and then coated with gold by a BAL-TEC SCD 005 sputter coater. The diameter of electrospun fibers was measured with image analyzer software (manual microstructure distance measurement). For each experiment, average fiber diameter, and distribution were determined from about 100 measurements of the random fibers. Electrical conductivity of the electrospun mats was measured by the standard four probe method after doping with HCl vapor.

TABLE 1 Factorial design of experiment.

Factor	Factor level
PANI Content(wt%)	10, 20, 30
Electrospinning temperature(°C)	25, 50, 75
Applied voltage(kV)	20, 25, 30

12.5 DISCUSSION AND RESULTS

At higher electrospinning temperature rate of solvent evaporation from the ejected jet increases significantly and a skin is formed on the surface of the jet which results collection of dry fiber with smooth surface. Presence of a thin, mechanically distinct polymer skin on the liquid jet during electrospinning has been discussed by Koombhongse et al. [27]. On the other hand, higher electrospinning temperature results higher degree of stretching and more uniform elongation of the ejected jet due to higher mobility and lower viscosity of the solution. Therefore, fibers with smaller diameters and narrower diameters distribution will be electrospun at higher electrospinning temperature (Figure 2).

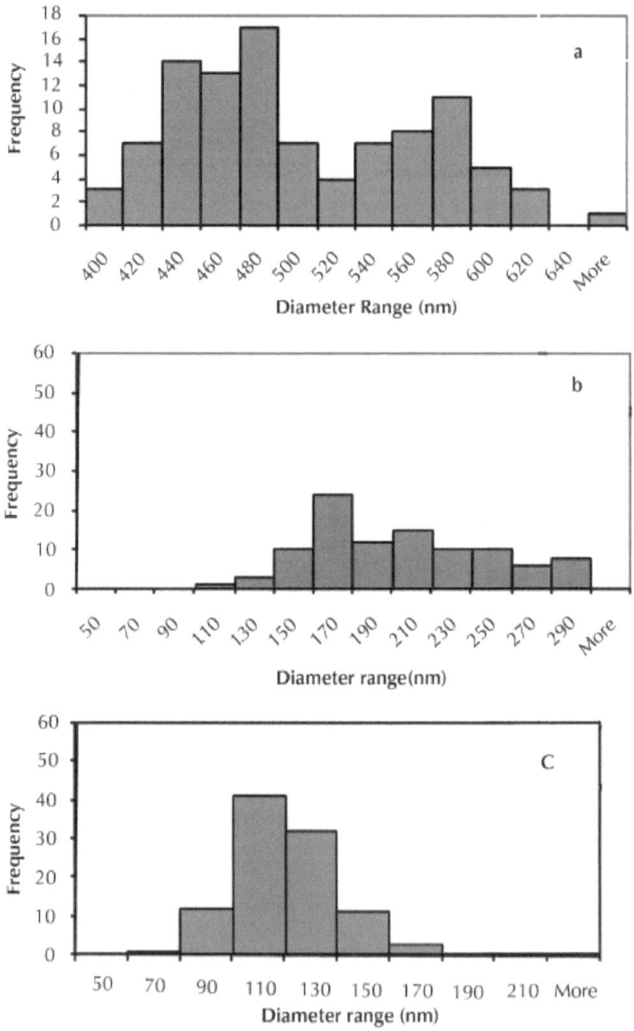

FIGURE 2 Distribution of fiber diameter electrospun at PANI content of 20%, applied voltage of 20 kV, spinning distance of 10 cm, and electrospinning temperature of (a) 25°C, (b) 50°C, and (c) 75°C.

In order to study the effects of applied voltage, the blend solutions were electrospun at various applied voltages and temperatures. It was obvious that the diameter of electrospun PANI/PAN fibers at 50°C decreased as the applied voltage increased. Similar results were observed for electrospun fibers at 25 and 70°C (results were not shown). The same results were found by Fenessey et al. [28] Ding et al. [29], and others [30-31] .But it is contradictory with the results obtained by Renker et al. [32] and Gu et al. [33] which found insignificant change of the diameter of electrospun fibers over the range of applied voltage. The inconsistency may be due to difference of experimental conditions. The flow rate of solution in experiments was maintained constant with the help of a syringe pump, while in the experiments of Gu et al. [33] the solution was brought down automatically by electrostatic force and hydrodynamic force of fluid. Therefore, the flow rate in their experiments was not constant. The diameter of fibers is combination results of flow rate and electrostatic force due to applied voltage. Increasing the applied voltage at constant flow rate increases the electrostatic force and creates smaller diameter fibers. But if the increasing of the applied voltage draws more solution out of the capillary, the fiber diameter would increased with increasing applied voltage as reported by Demir et al. [34]. In some reports [33] combination of increasing of applied voltage and flow rate resulted that the fiber diameter was not changed significantly with applied voltage.

The electrical conductivity of the electrospun mats at various PANI/PAN blend ratios was studies as well. As expected, electrical conductivity of the mats was found to increase with an increase in PANI content in the blends. We have found that the electrical conductivity of the mats increases sharply when the PANI content in the blends is less than 5% after which it will gradually reach to 10^{-1} S/cm at higher PANI content. This result is in agreement with the observations of Yang and co workers [35] which reported the electrical conductivity of PANI/PAN blend composites. Yang et al. [35] proposed the classical law of percolation theory, $\sigma(f) = c(f - f_p)^t$, where c is a constant, t is critical exponent of the equation, f is the volume fraction of the filler particle, and f_p is the volume fraction at percolation threshold. The study indicates that the conductivity of the mats follows the scaling law of percolation theory mentioned as shown in Equation (1) which results a value of 0.5 wt% of PANI for f_p. This value for the percolation threshold is much lower than that reported by Yang et al. [35] which may be due to the difference in the studied sample form. Their measurements were performed on the prepared films whereas measurements were performed on the nanofiber mats. It is worth noting that the classical percolation theory predicts a percolation threshold of $f_p = 0.16$ for conducting particles dispersed in an insulating matrix in three dimensions [35] which is in agreement of finding.

$$\sigma = 9\times10^{-7}(f - 0.5)^{3.91} \qquad R^2 = 0.99 \tag{1}$$

12.6 CONCLUSION

The electrospinning of PANI/PAN blend in NMP was processed and fibers with diameter ranging from 60 to 600 nm were obtained depending on electrospinning conditions.

The morphology of fibers was investigated at various blends ratios and electrospinning temperature. At 30% PANI content and 25°C fibers with average diameter of 164 nm were formed with beads (droplets of polymer over the woven mat) and not uniform morphology. At this condition solution viscosity and chain entanglements may not be enough, resulting in spraying of large droplets connected with very thin fibers. Averages of fibers diameters were decreased with PANI content in the solutions but PANI/PAN solution containing more than 30% PANI did not form a stable jet regardless of applied voltage and electrospinning temperature. For pure PANI solution, since the viscosity is too low to get stable drops and jets, we could not get the fibers. It was found that at 25°C fiber morphology was changed to beaded fibers when PANI content was higher than 20%. With increasing the electrospinning temperature, the morphology was changed from beaded fibers to uniform fibrous structure and the fiber diameter was also decreased from 500 to 100 nm when the electrospinning temperature changes from 25 to 75°C. The mean of fiber diameter is the smallest and the fiber diameter distribution is the narrowest for the electrospun fibers at 75°C. However some cracks are observed on the surface of the electrospun fibers. There was a slightly decrease in average fiber diameter with increasing applied voltage. It is concluded that the optimum condition for nanoscale and uniform PANI/PAN fiber formation is 20% PANI content and 50°C electrospinning temperature regardless of the applied voltage. The conductivity of the mats follows the scaling law of percolation theory which predicts a value of 0.5 wt% of PANI as percolation threshold for the blend of PANI/PAN.

KEYWORDS

- **Electrospinning**
- **Four probe method**
- **Molecular weight**
- **Polyaniline**
- **Polyaniline emeraldine base**

REFERENCES

1. Hall, N. Focus Article, *Chemical Communications 1*, (2003).
2. He, J. H. *Polymer*, **45**, 9067 (2004).
3. Kang, E. T., Neoha, K. G., and Tan, K. L. *Progress in polymer Science*, **23**, 277 (1998).
4. Carter, S. A., Angelopoulos, M., Karg, S., Brock, P. J., and Scott, J. C. *Applied Physics Letter*, **70**, 2067 (1997).
5. Kuo, C. T., Huang, R. L., and Weng, S. Z. *Synthetic Metals*, **88**, 101 (1997).
6. Adhikari, B. and Majumdar, S. *Progress in Polymer Science*, **29**, 699 (2004).
7. Trojanowicz, M. *Microchimica Acta*, **143**, 75 (2003).
8. Bakker, E. *Anal. Chem.*, **76**, 3285 (2004).
9. Lee, S. H., Ku, B. C., Wang, X., Samuelson, L. A., and Kumar, J. *Mat. Res. Soc. Symp. Pro.*, **708**, 403 (2002).
10. He, J. H., Wan, Y. Q., and Xu, L. *Chaos, Solitons and Fractals*, **33**, 26–37 (2007).
11. Wang, X. Y., Lee, S. H., Drew, C., Senecal, K. J., Kumar, J., and Samuelson, L. A. *Mat. Res. Soc. Symp. Pro.*, **708**, 397 (2002).

12. Wang, X. Y., Drew, C., Lee, S. H., Senecal, K. J., Kumar, J., and Samuelson, L. A. *Nano lett.*, **2**, 1273 (2004).
13. Huang, Z. M., Zhang, Y. Z., Kotaki, M., and Ramakrishna, S. *Composite Science and Technology*, **63**, 2223 (2003).
14. Xu, L., He, J. H., and Liu, Y. *International J. of Nonlinear Science and Numerical Simulation*, **8**(2), 199 (2007).
15. Reneker, D. H. and Chun, I. *Nanotechnology*, **7**, 216 (1996).
16. MacDiarmid, A. G., Jones Jr, W. E., Norris, I. D., Gao, J., Johnson Jr, A. T., Pinto, N. J., Hone, J., Han, B., Ko, F. K., Okuzaki, H., and Llaguno, M. *Synthetic Metals*, **119**, 27 (2001).
17. Norris, I. D., Shaker, M. M., Ko, F. K., and MacDiarmid, A. G. *Synthetic Metals*, **114**, 109 (2000).
18. Díaz-de León, M. J. Proceeding of the national conference on undergraduate research (NCUR). University of Kentucky, Lexington, Kentucky (March 15–17, 2001).
19. Zhou, Y., Freitag, M., Hone, J., Staii, C., Johnson, A. T., Pinto, N. J., and MacDiarmid, A. G. *Applied Physical Letter*, **83**, 18 (2003).
20. Formhals, A. US Patent # 1, 975, 504 (1934).
21. Teo, W. E. and Ramakrishna, S. *Nanotechnology*, **17**, R89 (2006).
22. Shenoy, S. L., Bates, W. D., Frisch, H. L., and Wnek, G. E. *Polymer*, **46**, 3372 (2005).
23. He, J. H., Wan, Y. Q., and Yu, J. Y. *Fibers and Polymers*, **9**(2), 140 (2008).
24. Deitzel, J. M., Kleinmeyer, J., Harris, D., and Beck Tan, N. C. *Polymer*, **42**, 261 (2001).
25. Theron, S. A., Zussman, E., and Yarin, A. L. *Polymer*, **45**, 2017 (2004).
26. Tan, S. H., Inai, R., Kotaki, M., and Ramakrishna, S. *Polymer*, **46**, 6128 (2005).
27. Koombhongse, S., Liu, W., and Renker, D. *J. of polymer science Part B polymer Physics*, **39**, 2598 (2001).
28. Fennessey, S. F. and Farris, R. J. *Polymer*, **45**, 4217 (2004).
29. Ding, B., Kim, H. Y., Lee, S. C., Shao, C. L., Lee, D. R., Park, S. L., et al. *J. of Polymer Science B*, **40**, 1261 (2002).
30. Mo, X. M., Xu, C. Y., Kotaki, M., and Ramakrishna, S. *Biomaterials*, **25**, 1883 (2004).
31. Katti, D. S., Robinson, K. W., Ko, F. K., and Laurencin, C. T. *J. of Biomedical Materials Research Part B*, **70**(B), 286 (2004).
32. Renker, D. H. and Chun, I. *Nanotechnology*, **7**, 216 (1996).
33. Gu, S. Y., Ren, J., and Vancso, G. J. *European Polymer J.*, **41**, 2559 (2005).
34. Demir, M. M., Yilgor, I., Yilgor, E., and Erman, B. *Polymer*, **43**, 3303 (2002).
35. Pen, W., Yang, S. L., Li, G., and Jiang, J. M. *European Polymer J.*, **41**, 2127 (2005).

13 Development of Conducting Polymers: (Part III)

A. K. Haghi and G. E. Zaikov

CONTENTS

13.1 INTRODUCTION

The objective of this chapter is to apply the electroless plating of Cu–Ni–P alloy on conducting polymers using hypophosphite as the reducing agent to the preparation of conductive fabrics and to investigate effects of operating parameters on the deposition rate. Electroless plating of Cu–Ni–P alloy on fabrics and effect of plating parameters on the properties of alloy coated fabrics as well as electromagnetic interference (EMI) shielding effectiveness were reported in detail. In this research conductive fabrics obtained with high shielding effectiveness.

Because of the high conductivity of copper, electroless copper plating is currently used to manufacture conductive fabrics. Electroless copper plating on fabrics has been studied by some researchers [1-6]. Electroless metal plating is a non electrolytic method of deposition from solution [7]. The electroless deposition method uses a catalytic redox reaction between metal ions and dissolved reduction agent[8-12]. This technique advantages, such as low cost, excellent conductivity, shielding effectiveness

(SE), easy formation of a continuous, and uniform coating on the surface of substrate with complex shapes. It can be performed at any step of the textile production, such as yarn, stock, fabric, and clothing [1].

Hence, chemical copper plating could be a unique process providing good potential for creation of fabrics with a metallic appearance and good handling characteristics. Revealing the performance of electroless plating of Cu-Ni-P alloy on cotton fabrics is an essential research area in textile finishing processing and for technological design. The main aim of this chapter is to explore the possibility of applying electroless plating of Cu-Ni-P alloy process onto cotton fabric. The fabrication and propertie of Cu-Ni-P alloy plated cotton fabric are investigated in accordance with standard testing methods.

13.2 EXPERIMENTAL

The fabrics (53 × 48 count/cm^2, 140 g/m^2, taffeta fabric) were used as substrate. The surface area of each specimen is 100 cm^2 .The electroless copper plating process was conducted by multistep processes: pre-cleaning, sensitisation, activation, electroless Cu-Ni-p alloy deposition, and post-treatment.

The fabric specimens (10 × 10 cm) were cleaned with non-ionic detergent (0.5 g/l) and $NaHCo_3$ (0.5 g/l) solution for 10 min prior to use. The samples then were rinsed in distilled water. Surface sensitization was conducted by immersion of the samples into an aqueous solution containing $SnCl_2$ and HCl. The specimens were then rinsed in deionized water and activated through immersion in an activator containing $PdCl_2$ and HCl. The substrates were then rinsed in a large volume of deionized water for 10 min to prevent contamination the plating bath. The electroless plating process carried out immediately after activation. Then all samples immersed in the electroless bath containing copper sulfate, nickel sulfate, sodium hypophosphite, sodium citrate, boric acid, and $K_4Fe(CN)_6$.

In the post-treatment stage, the Cu-Ni-P plated cotton fabric samples were rinsed with deionized water, ethylalcohol at home temperature for 20 min immediately after the metallising reaction of electroless Cu-Ni-P plating.Then dried in oven at 70°C.

The weights (g) of fabric specimens with the size of 100 × 100 mm square before and after treatment were measured by a weight meter (HR200, AND Ltd., Japan). The percentage for the weight change of the fabric is calculated in Equation (1).

$$I_W = \frac{W_f - W_0}{W_0} \times 100\% \qquad (1)$$

where, I_w is the percentage of increased weight, W_f is the final weight after treatment, and W_o is the original weight.

The thickness of fabric before and after treatment was measured by a fabric thickness tester (M034A, SDL Ltd., England) with a pressure of 10 g/cm^2. The percentage of thickness increment were calculated in accordance to Equations (2).

$$T_I = \frac{T_F - T_0}{T_0} \times 100\% \tag{2}$$

where, T_I is the percentage of thickness increment, T_f is the final thickness after treatment, and T_o is the original thickness.

A Bending Meter (003B, SDL Ltd., England) was employed to measure the degree of bending of the fabric in both warp and weft directions.The flexural rigidity of fabric samples expressed in Ncm is calculated in Equation (3).

$$G = W \times C^3 \tag{3}$$

where, G (Ncm) is the average flexural rigidity, W (N/cm^2) is the fabric mass per unit area, and C(cm) is the fabric bending length.

The dimensional changes of the fabrics were conducted to assess shrinkage in length for both warp and weft directions and tested accordance with (M003A, SDL Ltd., England) the stansdard testing method (BS EN 22313:1992). The degree of shrinkage in length expressed in percentage for both warp and weft directions were calculated in accordance to Equation (4).

$$D_c = \frac{D_f - D_0}{D_0} \times 100 \tag{4}$$

where, D_c is the average dimensional change of the treated swatch, D_o is the original dimension, and D_f is the final dimension after laundering.

Tensile properties and elongation at break load were measured with standard testing method ISO 13934-1:1999 using a Micro 250 tensile tester.

The color changing under different application conditions for two standard testing methods, namely, (1) ISO 105-C06:1994 (color fastness to domestic and commercial laundering), and (2) ISO 105-A02:1993 (color fastness to rubbing) were used for estimate.

The scanning electron microscope (SEM) (XL30 PHILIPS) was used to characterize the surface morphology of deposits. The wavelength-dispersive X-ray (WDX) analysis(3PC, Microspec Ltd., USA) was used to exist metallic particles over surface Cu-Ni-P alloy plated fabrics.

13.3 DISCUSSION AND RESULTS

13.3.1 Fabric Weight and Thickness

The change in weight and thickness of the untreated cotton and Cu-Ni-P alloy plated cotton fabrics are shown in Table 1.

TABLE 1 Weight and thickness of the untreated and Cu-Ni-P-plated cotton fabrics.

Specimen (10 × 10 cm)	Weight (g)	Thickness(mm)
Untreated cotton	2.76	0.4378
Cu-Ni-P plated cotton	3.72 (↑18.47%)	0.696(↑5.7%)

The results presented that the weight of chemically induced Cu-Ni-P plated cotton fabric was heavier than the untreated one. The measured increased percentages of weight was 18.47%.This confirmed that Cu-Ni-P alloy had clung on the surface of fabric effectively. In the case of thickness measurement, the cotton fabric exhibited a 5.7% increase after being subjected to metallization.

13.3.2 Fabric Bending Rigidity

Fabric bending rigidity is a fabric flexural behavior that is important for evaluating the handling of the fabric. The bending rigidity of the untreated cotton and Cu-Ni-P plated fabrics is shown in Table 2.

TABLE 2 Bending rigidity of the untreated and Cu-Ni-P-plated fabrics

Specimen	Bending (N·cm)	
	warp	weft
Untreated fabric	1	0.51
Cu-Ni-P plated fabric	1.17(↑11.39%)	0.66(↑30.95%)

The results proved that the chemical plating solutions had reacted with the original fabrics during the entire process of both acid sensitization and alkaline plating treatment. After electroless Cu-Ni-P alloy plating, the increase in bending rigidity level of the Cu-Ni-P plated cotton fabrics was estimated at 11.39% in warp direction and 30.95% in weft direction respectively. The result of bending indicated that the Cu-Ni-P plated cotton fabrics became stiffer handle than the untreated cotton fabric.

13.3.3 Fabric Shrinkage

The results for the fabric Shrinkage of the untreated fabric and Cu-Ni-P plated fabrics are shown in Table 3.

TABLE 3 Dimensional change of the untreated and Cu-Ni-P-plated fabrics.

Specimen	Shrinkage (%)	
	warp	weft
Untreated fabric	0	0
Cu-Ni-P plated fabric	–8	–13.3

The measured results demonstrated that the shrinkage level of the Cu-Ni-P plated cotton fabric was reduced by 8% in warp direction and 13.3% in weft direction respectively.

After the Cu-Ni-P plated, the copper particles could occupy the space between the fibres and hence more copper particles were adhered on the surface of fibre. Therefore, the surface friction in the yarns and fibres caused by the Cu-Ni-P particles could then be increased. When compared with the untreated fabric, the Cu-Ni-P plated fabrics shown a stable structure.

13.3.4 Fabric Tensile Strength and Elongation

The tensile strength and elongation of cotton fabrics was enhanced by the electroless Cu-Ni-P alloy plating process as shown in Table 4.

TABLE 4 Tensile strength and percentage of elongation at break load of the untreated and Cu-Ni-P- plated fabrics.

Specimen	Percentage of elongation (%)		Breaking load (N)	
	warp	weft	warp	weft
Untreated fabric	6.12	6.05	188.1	174.97
Cu-Ni-P plated fabric	6.98(↑12.5%)	6.52(↑7.8%)	241.5(↑28.4%)	237.3(↑35.62%)

The metallized fabrics had a higher breaking load with a 28.44% increase in warp direction and a 35.62% increase in weft direction than the untreated fabric. This was due to the fact that more force was required to pull the additional metal layer coating.

The results of elongation at break were 12.5% increase in warp direction and 7.8% increase in weft direction, indicating that the Cu-Ni-P plated fabric encountered little change when compared with the untreated fabric. This confirmed that with the metallizing treatment, the specimens plated with metal particles was demonstrated a higher frictional force of fibers. In addition, the deposited metal particles which developed a linkage force to hamper the movement caused by the applied load.

13.3.5 Color Change Assessment

The results of evaluation of color change under different application conditions, washing, and rubbing are shown in Table 5.

TABLE 5 Washing and rubbing fastness of the untreated and Cu-Ni-P Fabrics plated cotton fabrics.

Specimen	Washing	Rubbing	
		Dry	Wet
Cu-Ni-P plated cotton	5	4-5	3-4

The results of the washing for the Cu-Ni-P plated cotton fabric was grade 5 in color change. This confirms that the copper particles had good performance during washing. The result of the rubbing fastness is shown in Table 5. According to the test result, under dry rubbing condition, the degree of staining was recorded to be grade 4–5, and the wet rubbing fastness showed grade 3–4 in color chang. This results showed that the dry rubbing fastness had a lower color change in comparison with the wet crocking fastness. In view of the overall results, the rubbing fastness of the Cu-Ni-P plated fabric was relatively good when compared with the commercial standard.

13.3.6 Shielding Effectiveness (SE)

The electromagnetic shielding means that the energy of electromagnetic radiation is attenuated by reflection or absorption of an electromagnetic shielding material, which is one of the effective methods to realize electromagnetic compatibility .The unit of electromagnetic interference shielding effectiveness (EMISE) is given in decibels (dB). The EMISE value was calculated from the ratio of the incident to transmitted power of the electromagnetic wave in the following equation:

$$SE = 10\log\left|\frac{P_1}{P_2}\right| = 20\left|\frac{E_1}{E_2}\right| \quad (5)$$

where, P_1 (E_1) and P_2 (E_2) are the incident power (incident electric field) and the transmitted power (transmitted electric field), respectively. This study indicates the SE of the uncoated and copper-coated fabrics with 1 ppm $K_4Fe(CN)_6$. As a result, the SE of cotton fabrics was almost zero at the frequencies 50 MHz to 1.5 GHz. However, SE of copper-coated cotton fabric was above 90 dB and the tendency of SE kept similarity at the frequencies 50 MHz to 1.5 GHz. The copper-coated cotton fabric has a practical usage for many EMI shielding application requirements.

13.4 CONCLUSION

In this study, electroless plating of Cu-Ni-P alloy process onto fabrics was demonstrated. Both uncoated and Cu-coated cotton fabrics were evaluated with measurement weight change, fabric thickness, bending rigidity, fabric shrinkage, tensile strength, percentage of elongation at break load, and color change assessment. The results showed significant increase in weight and thickness of chemically plated fabric. Coated samples showed better properties and stable structure and with uniformly distributed metal particles.

KEYWORDS

- **Cotton fabric**
- **Deposition rate**
- **Electroless plating of Cu-Ni-P alloy**
- **Surface morphology**

REFERENCES

1. Guo, R. H., Jiang, S. Q., Yuen, C. W. M., and Ng, M. C. F. *An alternative process for electroless copper plating on polyester fabric*. (2008).

2. Xueping, G., Yating, W., Lei, L., Bin, Sh., and Wenbin, H. Electroless plating of Cu–Ni–P alloy on PET fabrics and effect of plating parameters on the properties of conductive fabrics. *Journal of Alloys and Compounds*, **455**, 308–313 (2008).
3. Li, J., Hayden, H., and Kohl, P. A. The influence of 2,2'-dipyridyl on non-formaldehyde electroless copper plating. *Electrochim. Acta*, **49**, 1789–1795 (2004).
4. Larhzil, H., Cisse, M., Touir, R., Ebn Touhami, M., and Cherkaoui, M. Electrochemical and SEM investigations of the influence of gluconate on the electroless deposition of Ni–Cu–P alloys. *Electrochimica Acta*, **53**, 622–628 (2007).
5. Gaudiello, J. G., Ballard, G. L. Mechanistic insights into metal-mediated electroless copper plating Employing hypophosphite as a reducing agent. *IBM J. RES. DEVELOP*, **37**(2), (1993).
6. Xueping, G., Yating, W., Lei, L., Bin, Sh., and Wenbin, H. Electroless copper plating on PET fabrics using hypophosphite as reducing agent. *Surface & Coatings Technology*, **201**, 7018–702 (2007).
7. Han, E. G., Kim, E. A., and Oh, K. W. Electromagnetic interference shilding effectiveness of electroless Cu-platted PET fabrics. *Synth. Met.*, **123**, 469–476 (2001).
8. Jiang, S. Q. and Guo R. H. Effect of Polyester Fabric through Electroless Ni-P Plating. *Fibers and Polymers*, **9**(6), 755–760 (2008).
9. Djokic, S. S. *Fundamental Aspects of Electrometallurgy*: Chapter 10, Metal Deposition without an External Current. K. I. Popov, S. S. Djokic, and B. N. Grgur (Eds.). Kluwer Academic Publishers, New York, pp. 249–270 (2002).
10. O. Mallory and B. J. Hajdu (Eds.) *Eleolesctrs Plating: Fundamentals and Applications*. Noyes Publication, New York (1990).
11. Agarwala, R. C., and Agarwala, V. *Electroless alloy/composite coatings: a review*. Sadhana, **28**, 475–493 (2003).
12. Paunovic, M. Electrochemical Aspects of Electroless Deposition of Metals. *Plating*, **51**, 1161–1167 (1968).

14 Development of Conducting Polymers: (Part IV)

A. K. Haghi and G. E. Zaikov

CONTENTS

14.1 INTRODUCTION

The conducting polymers because of their electronic, magnetic, and optical properties are an attractive class of materials for variety of advanced technologies [1-5]. Among conducting polymers, polyaniline (PANI) have been extensively studied due to its good environmental stability, electrical properties, and inexpensive monomer. In this chapter a new synthetic strategy to obtain ring substituted poly(hydroxyaniline) and poly(iodoaniline), using poly (aniline boronic acid) as a precursor for substitution. Herein, we report the synthesis of propyl thiol ring substituted PANI by reaction of 3-chloropropyl thiol and ferric chloride with dimethylsulfoxide (DMSO) solution of emeraldine base form of PANI. The 3-chloropropyl thiol ring substitution of PANI was also carried out on the surface glassy carbon (GC) electrode.

14.2 EXPERIMENTAL

14.2.1 Material and Reagents

The 3-chloropropyl thiol, aniline, ammonium persulfate, acetonitrile, DMSO, tetrahydrofuran (THF), and ferric chloride were purchased from Aldrich Chemical Inc. Hydrochloric acid was purchased from Fisher Scientific.

14.2.2 Instrumental Setup

The GC electrodes (3 mm diameter) were purchased from bioanalytical science. The cyclic voltammetry was performed with a bioanalytical system (BAS) potentiostat (Model 100). In the voltammetric experiments, a three-electrode configuration was used, including Ag/AgCl reference electrode and a platinum wire counter electrode. The reflectance fourier transform infrared (FTIR) spectra of polymers were obtained using Nicolet NEXUS 870 FTIR instrument. The UV-vis. spectra use an Agilent 8453 spectrophotometer.

14.2.3 Reaction of 3-chloropropyl thiol and Ferric Chloride with PANI Deposited on the GC Electrode

The oxidative polymerization of aniline was performed to produce PANI, when 40 mM of aniline was dissolved in 25 ml 0.5M aqueous hydrochloric acid solution. The potential of the GC electrode was scanned between 0.4 and 1.1 V versus Ag/AgCl at a scan rate 100 mV/s, the polymerization was stopped at 0.4 V when the charge passed from the reduction of deposited polymer reached to 0.65 V. The PANI film had a green color. The film was washed with distilled water, then with THF. The GC electrode with film was immersed in a large glass tube with THF solution contained 0.01 g $FeCl_3$ and 0.5 ml of 3-chloropropyl thiol the top of tube was closed to prevent evaporation of the solvent. The reaction was carried out at 60°C for 24 hr in a temperature controlled mineral oil bath. The cyclic voltammogram of the film after reaction is similar to those 1, 2, 4 trisubstituted PANI reported by Freund.

14.2.4 Reaction of 3-chloropropyl thiol and Ferric Chloride with PANI

The emeraldine base form of PANI (compound 1) was prepared by chemical oxidation method described by MacDiarmid and co-workers [33]. The PANI was extracted with THF until the extract was colorless. The dried PANI (0.25 g) reacted with 3-chloropropyl thiol (0.5 ml) and ferric chloride (0.01 g) in anhydrous dimethyl sulfoide at 81°C for 24 hr. The black solution (propyl thiol substituted PANI) was precipitated with 1M HCl, a dark green to black precipitate was formed the filtrate was greenish blue; the precipitate was washed with 1M HCl, then with acetonitrile. The precipitate (propyl thiol substituted PANI) was dried under continuous vacuum for 48 hr.

The HCl doped Propyl thiol polyaniline was sent for chemical analysis to Guelph Chemical Laboratories LTD (Guelph, Ontario, Canada). The result of chemical analysis suggests that substitution occurred on all benzenoid rings in the polymer (compound 2).

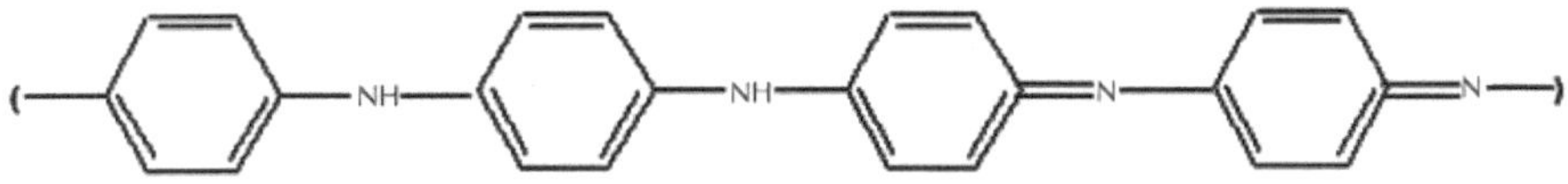

compound 1

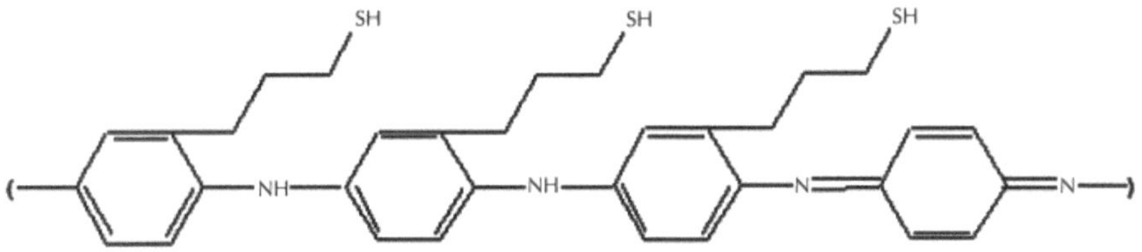

compound 2

14.2.5 Characterization

The propyl thiol substituted PANI was studied by reflectance FTIR using Nicolet NEXUS 870. The polymer was treated with 1M aqueous NaOH solution to remove the doped HCl from the polymer. The polymer was washed with water and then, it was dried under continuous vacuum for 24 hr. The FTIR reflectance spectrum of sodium polyaniline propyl thiol sulfide (NaPAPS) was obtained and compared with the FTIR reflectance spectrum of emeraldine base.

The UV-vis. spectrum of DMSO solution of propyl thiol polyaniline was studied using an Agilent 8453 spectrophotometer.

The cyclic voltammogram of Propyl thiol polyaniline was obtained in 0.5M HCl solution using BAS Instrument Model 100, a thin film was smeared on indium tin oxide (ITO) glass slide, the working electrode, a platinum wire counter electrode, a platinum wire counter electrode, and Ag/AgCl reference electrode.

14.3 DISCUSSION AND RESULTS

The effect of substituent on solubility, therefore processibility of PANI has been a subject of interest of scientists and engineers for many years. To develop a new synthetic approach to obtain substituted PANI, in present investigation the Friedel–Craft synthetic method was used for production of propyl thiol substituted PANI.

The PANI was prepared by both direct oxidation of aniline using ammonium persulfate and electrochemical oxidation of aniline on GC electrode.

The FTIR spectrum of sodium salt of propyl thiol substituted PANI, Figure 1 shows that the ratio of absorption intensity at 1594 cm^{-1} due to the quinoid ring to that at 1506 cm^{-1} due to that benzenoid ring is 1:3 [34, 35], as it presented in the compound 1. The C-S stretching band of propyl thiol appeared at 775 cm^{-1} [36, 37]. The aliphatic C-H stretch at 2947cm^{-1}, 2915 cm^{-1}, and symmetrical stretching versus CH_2 at 2847 cm^{-1} confirms the propyl thiol substitution of PANI.

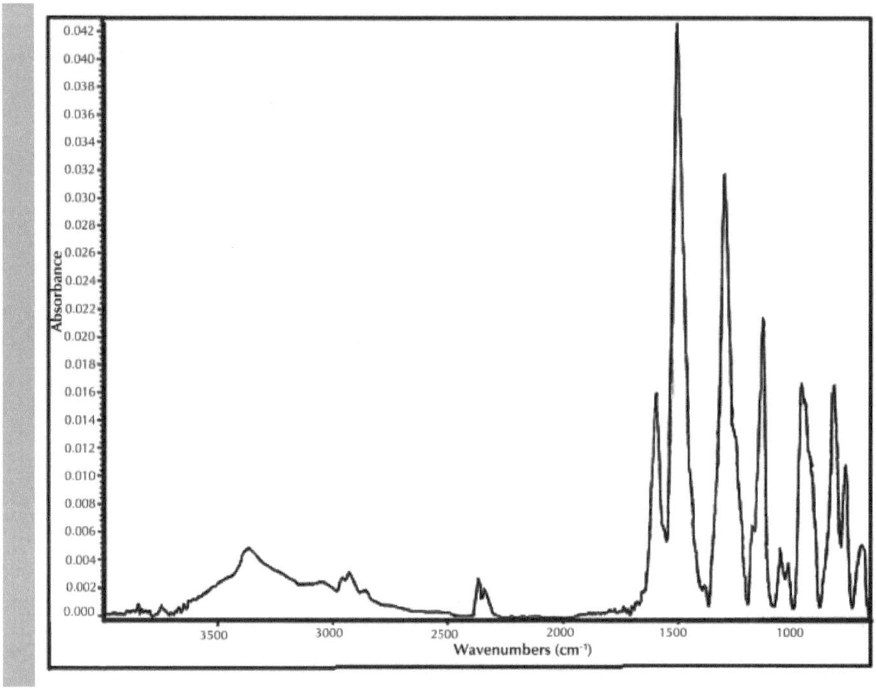

FIGURE 1 The refractive FTIR of sodium propyl sulfide substituted PANI.

TABLE 1 Chemical analysis of o-propyl thiol substituted PANI.

	carbon	hydrogen	nitrogen	Sulfur
Theoretical values	60	5.5	8.5	14.6
Corrected chemical analysis values	59.8	5.4	8.3	15
Observed chemical analysis* values of substituted polymer	54.50	4.82	7.38	13.43

* The propyl thiol substituted PANI was precipitated with 1M solution of HCl then it was washed with HCl solution then with acetonitrile. The propyl thiol substituted PANI contained two moles of HCl.

The chemical analysis of HCl doped propyl thiol polyaniline Table 1, perfectly suggests that substitution occurred on all benzenoid rings in the polymer. Three out of four rings in PANI were substituted, it makes the degree of substitution 75%, while the degree of substitution of electrochemically prepared ring substituted dialkylamine is 25% [34]. Electrophilic substitution of chloroalkyl thiole in the presence of ferric chloride makes it possible to prepare any type of alkyl thiole, alkyl amine, and other substitution with a good degree of substitution.

14.4 CONCLUSION

The production of substituted PANI(s) seems to be straight forward and it can be generated *via* oxidative polymerization of corresponding monomer. However, in many cases the desired substituted polymer is hard to obtain. In present investigation PANI(s) have been prepared chemically by direct oxidation of aniline and electrochemical oxidation of aniline on GC electrode. The substitution reaction of polymer with 3-chloropropyl thiol and ferric chloride occurred *via* Friedel-Craft mechanism. Although for bulk chemically prepared polymer this method is restricted due to low to moderate solubility of polymer in the proper solvent, but this technique works perfectly for polymers prepared on the surface of electrode or prepared on the surface of conducting glass slide. We suggest Friedel–Craft method for synthesis is a good strategy to overcome the mentioned problems for preparation of substituted polymers.

KEYWORDS

- **Conducting polymers**
- **Fourier transform infrared spectra**
- **Glassy carbon electrodes**
- **Polyaniline**
- **Propyl thiol**

REFERENCES

1. Schultze, J. W. and Karabulut, H. *Electrochim. Acta*, **50**, 1739 (2005).
2. Skoheim, T. A., Elesenbaumer, R. L., and Renolds, J. R. *Hand book of conducting polymers*, 2nd ed., Marcel Dekker, New York (1998).
3. Gustafsson, G., Cao, Y., Treacy, G. M., Klavetter, F., Colaneri, N., and Heeger, A. G. J. *Nature*, **357**, 477 (1992).
4. Jaeger, W. H., Inganas, O., and Lundstrom, I. *Science*, **288**, 2335 (2000).
5. Wang, W., Sotzing, G. A., and Weiss, R. A. *Chem. Mater*., **15**, 375 (2003).
6. MacDiarmid, A. G., Mu, S. L., Somarisi, M. L. D., and Wu, W. *Mol. Crist. Liq. Crist.*, **121**, 187 (1985).
7. Nakajima, T. and Kawagoe, K. *Synth. Met.*, **28**(C), 629 (1989).
8. Shoji, E. *Chem. Sensor*, **21**(4), 120 (2005).
9. Zeng, K., Tachikawa, H., and Zhu, Z. *Anal. Chem*., **72**, 2211 (2000).
10. English, J. T., Deore, B. A., and Freund, M. S. *Sensor Actuat B Chem. B*, **115**(2), 666 (2006).
11. Michelotti, F., Morelli, M., Cataldo, F., Petrocco, G., and Bertolotti, M. *SPIE Proceedings Series*, **2042**, 186 (1994).
12. Paul, E. W., Ricco, A. J., and Wrighton, M. S. *J. Phys. Chem.*, **89**, 1441 (1985).
13. Chao, S. and Wrighton, M. S. *J. Am. Chem. Soc*., **109**, 6627 (1987).
14. Kobayashi, T., Yoneyama, H., and Tamaru, H. *J. Electroanal. Chem*., **209**, 227 (1986).
15. Helay, F. M., Darwich, W. M., and Elghaffar, M. A. *Poly. Degrad. Stab.*, **64**, 251 (1999).
16. Krinichnyi, V. I., Yeremenko, O. N., Rukhman, G. G., Letuchii, Y. A., and Geskin, V. M. *Polym. Sci. USSR*, **31**, 1819 (1989).
17. Gospodinova, N., Mokreva, P., Tsanov, T., and Terlemezyan, L. *Polymer*, **38**, 743 (1997).
18. Banerjee, P. *Eur. Polym. J.*, **34**, 1557 (1998).
19. Sobczak, J. W., Kosinski, A., Biliniski, A., Pielaszek, J., and Palczewska, W. *Adv. Mater. Opt. Electron*, **8**, 213 (1998).

20. Athawale, A. A. and Kulkarni, M. V. *Sensor Actuator B*, **67**, 173 (2000).
21. Lin, C. W., Hwang, B. J., and Lee, C. R. *Mater. Chem. Phys.*, **55**, 139 (1998).
22. Wood, A. S. *Modern plastic Int.*, (August 3, 1991).
23. Epstein, A. J. and Yue, J. US Patent No. 5137991 (1992).
24. Bissessur, R. and White, W. *Mat. Chem. Phys.*, **99**, 214 (2006).
25. Kitani, A., Satoguchi, K., Tang, H. Q., Ito, S., and Sasaki, K. *Synth. Met.*, **69**, 131 (1995).
26. Lin, D. S. and Yang, S. M. *Synth. Met.*, **119**, 14 (2001).
27. Cattarin, S., Doubova, L., Mengoli, G., and Zotti, G. *Electrochemica. Acta*, **33**, 1077 (1988).
28. Ranger, M. and Leclerc, M. *Synth. Met.*, **84**, 85 (1997).
29. Sazou, D. *Synth. Met.*, **118**, 133 (2001).
30. Planes, G. A., Morales, G. M., Miras, M. C., and Barbero, C. *Synth. Met.*, **97**, 223 (1998).
31. Pringsheim, E., Terpetsching, E., and Wolfbeis, O. S. *Anal. Chim. Acta*, **357**, 247 (1997).
32. Shoji, E. and Freund, M. S. *Langmuir*, **17**, 7183 (2001).
33. Chiang, J. C. and MacDiarmid, A. G. *Synth. Met.*, **13**, 193 (1986).
34. Wudl, F., Angus, R. O., Allemand, P. M., Vachon, D. J., Norwak, M., Liu, Z. X., and Heeger, A. J. *J. Am. Chem. Soc.*, **109**, 3677 (1987).
35. Tang, H. T., Kitani, A., Yamashita, T., and Ito, S. *Synth. Met.*, **96**, 43 (1998).
36. Nakanishi, K. and Solomon, P. H. *Infrared Absorption Spectroscopy*, 2nd ed., Nankodo, Tokyo, p. 50 (1977).
37. Silverstein, R. M., Bassler, G. C., and Morrill, T. C. *Spectrometric Identification of Organic compounds*, 4th ed., John Wiley and Sons, p. 131 (1981).

15 Development of Conducting Polymers: (Part V)

A. K. Haghi and G. E. Zaikov

CONTENTS

15.1 INTRODUCTION

As mentioned in studies, conductive polymers such as polypyrrole (PPy), polyacetylene, and so on continue to be the focus of active research in diverse fields including electronics, energy storage catalysis, chemical sensing, and biochemistry. Among the conducting polymers, the PPy can be coated onto textile surfaces using either electrochemical or chemical oxidation of pyrrole in the presence of the fabric. It has been shown that non-conducting fibers such as cotton, silk, and wool or synthesis fibers become conductive when they are coated with PPy. It has been found that the *in situ* polymerization method is able to produce conductive PPy coated textiles, such as nylon, Lycra, and Taffeta polyester fabrics suitable for such a purpose.

In the current chapter, preparation of electrically conductive polyvinyl acetate (PVA) nanofiber using vapor phase chemical polymerization of pyrrole onto the surface of the nanofibers are investigated. The PVA was selected for this study because of its excellent chemical resistance and physical properties. The PVA is highly biocompatible and nontoxic.

15.2 EXPERIMENTAL

The PVA (Mw = 78,000) was purchased from Merck. Reagent grade pyrrole was obtained from Aldrich and distilled prior to use. All the other reagents were laboratory grades and used as received.

In order to prepare electrospinning solution, a 10% (w/w) PVA aqueous solution was first prepared. The required amount of oxidant (ammonium, persulfate, or $FeCl_3$) was then added to the prepared PVA solution and stirred using mechanical stirrer to form a homogenous solution. The mixtures containing various PVA/oxidant weight ratios were prepared and used for electrospinning.

In the electrospinning process, a high electric potential (Gamma High voltage) was applied to a droplet of PVA/oxidant solution at the tip of a syringe needle. A syringe pump (New Eva Pump System Inc, USA) was used to form a constant amount of solution on the tip. The output of the injection pump was 20 μl/min. A charged jet is formed and ejected in the direction of the applied field. The electrospun nanofibers were collected on a target alluminum foil which was placed at a distance of 10 cm from the syringe tip. A high voltage in the range from 10 to 20 kV was applied to the droplet of solution at the tip of the syringe and the best condition for the stable electrospinning was selected which was 15 kV.

The PPy was deposited on the PVA nanofibers during electrospinning of the nanofibers. Polymerization of the PPy was performed in the vapor phase at various conditions such as oxidant type and oxidant content in the electrospinning solution. For this purpose, a chamber was designed and electrospinning was done in this chamber (Figure 1).

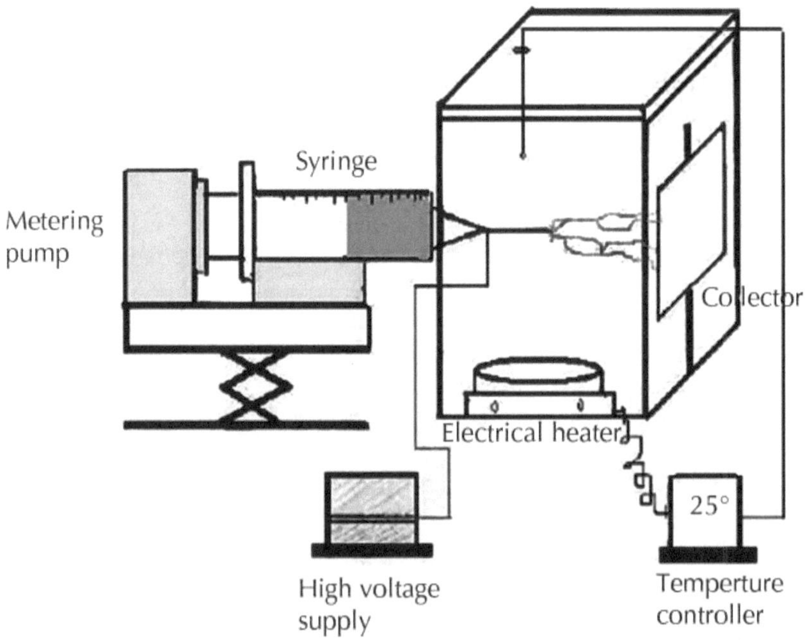

FIGURE 1 Schematic diagram of the electrospinning apparatus.

The fiber formation and morphology of the coated nanofibers were determined using a scanning electron microscope (SEM), Philips XL-30. A small section of the produced web was placed on SEM sample holder and coated with gold (BAL-TEC SCD 005 sputter coater). Electrical conductivity of the coated mats was determined by employing the standard four probe technique.

15.3 DISCUSSION AND RESULTS

Preliminary studies were performed to find optimum condition for electrospinning of PVA solution. The SEM photomicrographs of electrospun PVA nanofibers from 10% solution showed that there was a uniform nanofibers with diameter ranging from 150 to 400 nm, average diameter of 350 nm. As the polymer concentration increased, a mixture of beads and fibers were formed. Therefore, PVA concentration of 10% was used for preparation of electrospun nanofibers. A series of mats were prepared from the 10% PVA solution in mixture with various amount of $FeCl_3$ or ammonium persulfate ($(NH_4)_2S_2O_8$) as oxidant at the 15 KV constant electric field.

Figure 2 shows the SEM photomicrograph of PPy coated electrospun nanofiber using ammonium persulfate as oxidant at PVA/oxidant weight ratio of 3/1. It is observed that uniform layer of PPy with accumulations of PPy are deposited on the nanofibers surfaces. Average fiber diameter of uncoated nanofibers was 350 nm.

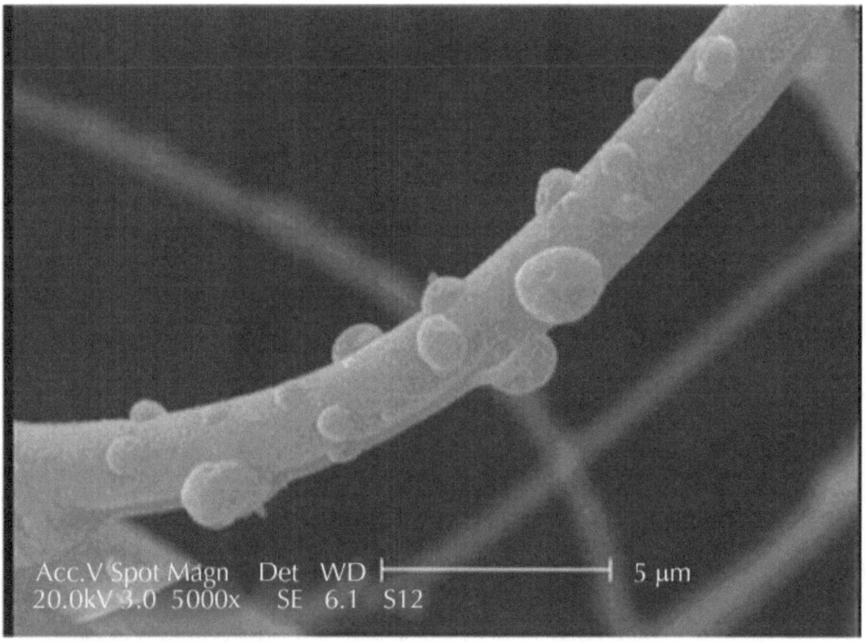

FIGURE 2 *(Continued)*

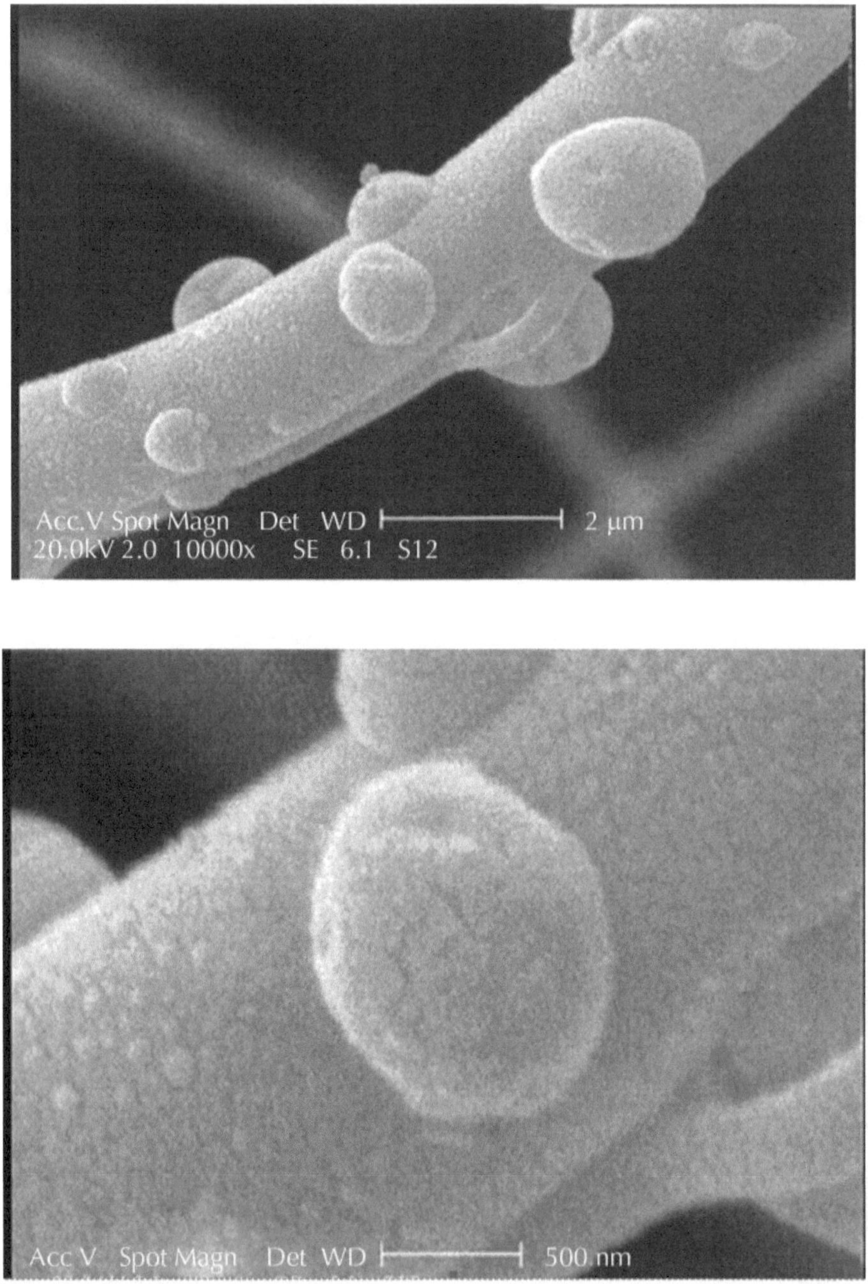

FIGURE 2 The SEM photomicrograph of PPy coated electrospun PVA nanofiber using ammonium persulfate as oxidant (PVA/oxidant = 3/1).

After coating process, average fiber diameter raised to 1,300 nm. It means that a layer of PPy with thickness of approximately 1,000 nm was coated onto the nanofibers. By changing the PVA/oxidant weight ratio to 2/1 and 1/1 (increase in oxidant content), deposition of the PPy onto the fiber was reduced and irregular fibers were obtained as shown in Figure 3. Electrical conductivity of the coated nanofiber mats depends upon the PVA/oxidant weight ratio. It seems that by suitable selection of the PVA/oxidant ratio and suitable coating of nanofibers increase in electrical conductivity can be achieved. Measurement of the electrical conductivity of the coated mats shows electrical conductivity in the range of 10^{-4}–10^{-1} S/cm depending on the chemical coating conditions such as amount of the oxidant in the PVA solution. As discussed on SEM photomicrograph, PVA/oxidant ration of 3/1 is the optimum condition for preparation uniform coated nanofibers with higher electrical conductivity.

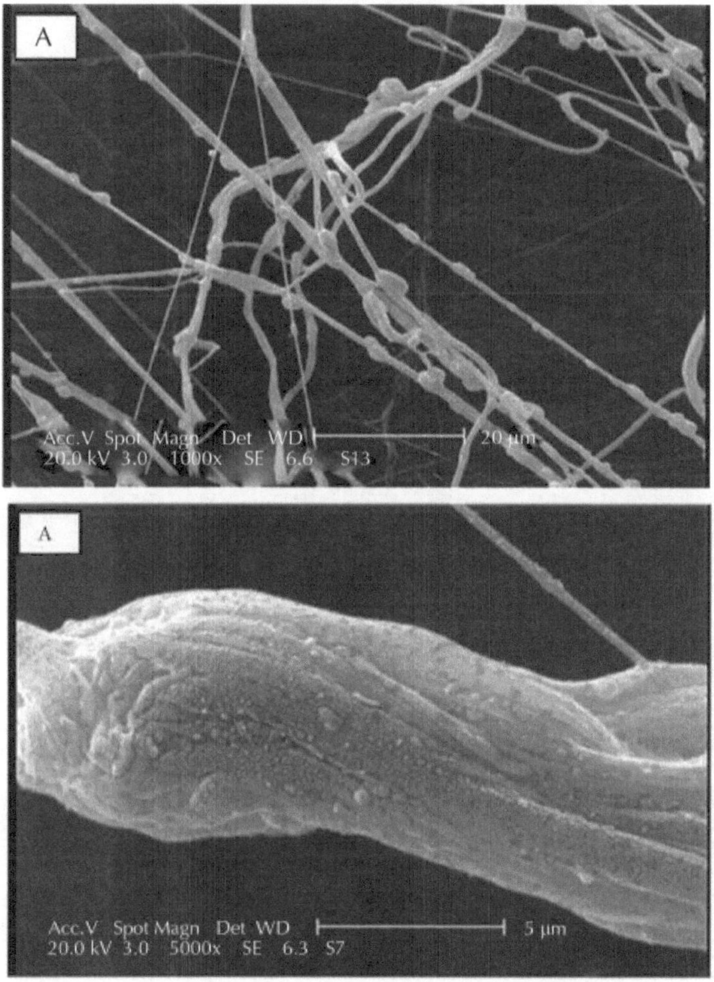

FIGURE 3 *(Continued)*

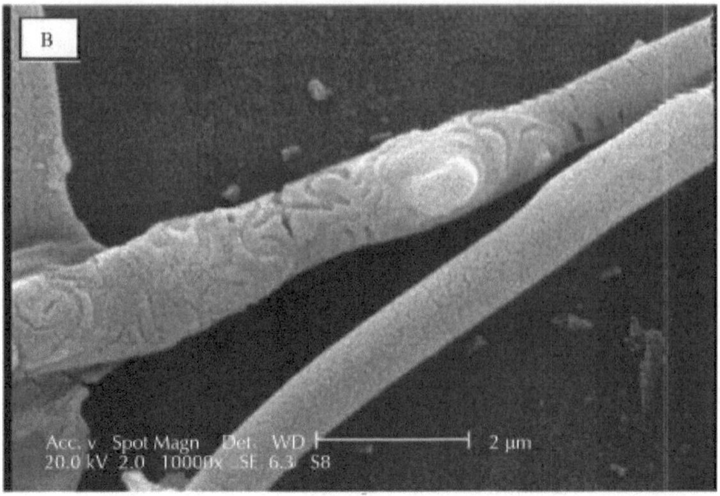

FIGURE 3 The SEM photomicrograph of PPy coated electrospun PVA nanofiber using as oxidant ((A) PVA/oxidant = 1/1, (B) PVA/oxidant = 2/1)

Figure 4 shows the SEM photomicrograph of PPy coated electrospun nanofiber using $FeCl_3$ as oxidant. In compare to ammonium persulfate coated nanofibers, there is not a distinct sign of PPy coating onto the fiber surface. The resulted webs are not conductive which confirms that the coating of nanofibers using $FeCl_3$ was not successful.

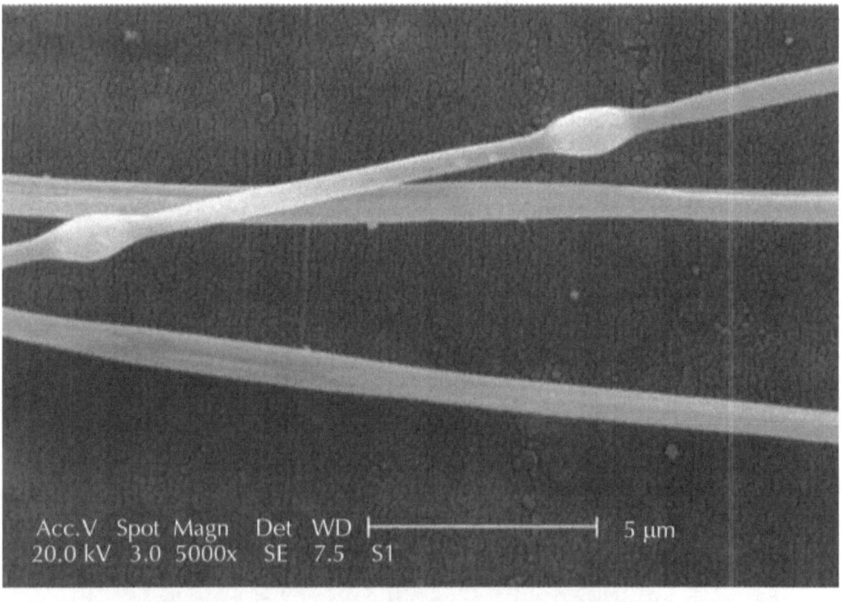

FIGURE 4 The SEM photomicrograph of PPy coated electrospun PVA nanofiber using $FeCl_3$ as oxidant (PVA/oxidant = 1/1).

For comparison, Figure 5 shows the SEM photomicrograph of PPy coated electrospun Polyacrylonitrile nanofiber using chemical *in situ* polymerization [12]. In this case, a uniform layer of PPy is deposited on the nanofibers surfaces. Average fiber diameter for uncoated nanofibers was 230 nm. After coating process, average fiber diameter increased to 350 nm. It means that a layer of PPy with the thickness of approximately 120 nm was coated onto the PAN nanofibers. Measurement of the electrical conductivity of the coated mats showed that electrical conductivity in the range of 10^{-3}–1 S/cm depending on the chemical coating conditions was achieved. This is at least one order of magnitude higher than those achieved through vapor phase polymerization reported in this work.

FIGURE 5 The SEM photomicrograph of PPy coated electrospun PAN nanofiber using solution polymerization [12].

15.4 CONCLUSION

The vapor phase chemical deposition of PPy on the surface of the electrospun PVA nanofibers was successfully performed. Morphology of the nanofibers and successful deposition of the PPy onto the nanofibers is governed by the oxidant type and its concentration in the electrospinning solution. At the best studied condition, a layer of PPy with the thickness of 1,000 nm is formed onto the PVA nanofibers surface using ammonium persulfate as oxidant. In contrast, surface coating was not observed in the case of $FeCl_3$ as oxidant. The electrical conductivity of the coated nanofiber mats depends on the coating conditions and reaches to 10^{-1} S/cm. Large surface area of the sensor material with the electrochemical activity is one of the most important properties for sensor application. Since, it is obvious that the surface area of PPy coated nonwoven web is much greater than that of the film with same volume or weight which makes the PPy coated nonwoven web potentially being as sensors.

REFERENCES

1. Wallace, G. G., Spinks, G. M., Kane-maguire, L. A. P., and Teasdale, P. R. *Conductive electroactive polymers* 3rd Ed., C.R.C., USA (2003).
2. Malinauskas, A. Chemical deposition of conducting polymers. *Polymer*, **42**, 3957–3972 (2001).
3. Rossi, D. D., Sanata, A. D., and Mazzoldi, A. Dressware wearable hardware. *Mater. Sci. Eng.*, **C7**, 31–37 (1999).
4. Kim, H. K., Kim, M. S., Chun, et al. Characterization of Electrically Conductive Polymer-Coated Textiles. *Mol. Cryst. Liq. cryst.*, **405**, 161–169 (2003).
5. Li, Y., Cheng, X. Y., Leung, M. Y., Sang, J. T., Tao, K. M., and Yuen, M. C. W. A flexible strain sensor from polypyrrole-coated fabrics. *Synth. Met.*, **155**, 89–94 (2005).
6. He, J. H., Wan, Y. Q., and Xu, L. Nano-effects, quantum-like properties in electrospun nanofibers. *Chaos, Solitons, and Fractals*, **33**, 26–37 (2007).
7. Wang, X. Y., Drew, C., Lee, S. H., Senecal, K. J., Kumar, J., and Samuelson, L. A. Electrospun nanofibers membranes for highly sensitive optical sensors. *Nano lett.*, **2**(11), 1273–1275 (2002).
8. Huang, Z. M., Zhang, Y. Z., Kotaki, M., and Ramakrishna, S. A review on polymer nanofibers by electrospinning and their applications in nanocomposites. *Composite Science and Technology*, **63**, 2223–2253 (2003).
9. Deitzel, J. M., Kleinmeyer, J., Harris, D., and Beck Tan, N. C. The effect of processing variables on the morphology of electrospun nanofibers and textiles. *Polymer*, **42**, 261–272 (2001).
10. Theron, S. A., Zussman, E., and Yarin, A. L. Experimental investigation of the governing parameters in the electrospinning of polymer solutions. *Polymer*, **45**, 2017–2030 (2004).
11. Tan, S. H., Inai, R., Kotaki, M., and Ramakrishna, S. Systematic parameter study for ultra-fine fiber fabrication via electrospinning process. *Polymer*, **46**, 6128–6134 (2005).
12. Mirbaha, H., Nouri. M, and Ghamgosar A. *Polypyrrol Coated Polyacrylonitril Electrospun Nanofibers*. AUTEX2008, biella, Italy (2008).

16 Development of Conducting Polymers: (Part VI)

A. K. Haghi and G. E. Zaikov

CONTENTS

16.1 INTRODUCTION

Electrospinning of polymers (for example, vibration-electrospinning, magneto-electrospinning, and bubble-electrospinning) is a simple and relatively inexpensive mean of manufacturing high volume production of very thin fibers (more typically 100 nm to 1 micron) and lengths up to kilometers from a vast variety of materials including polymers, composites, and ceramics [1, 2]. Electrospinning technology was first developed and patented by Formhals [3] in the 1930s and a few years later the actual developments were triggered by Reneker and co-workers [4]. To satisfy the increasing needs for the refined nanosize hybrid fibers based on commercial polymers, various electrospinning techniques have been investigated and developed [5]. Presently, there are two standard electrospinning setups, vertical and horizontal. With the development of this technology, several researchers have developed more intricate systems that can fabricate more complex nanofibrous structures in a more controlled and efficient style [6]. The unique properties of nanofibers are extraordinarily high surface area per unit mass, very high porosity, tunable pore size, tunable surface properties, layer thinness, high permeability, low basic weight, ability to retain electrostatic charges, and cost effectiveness [2]. In this method nanofibers produce by solidification of a polymer solution stretched by an electric field [7-9] which can be applied in different areas including wound dressing, drug, or gene delivery vehicles, biosensors, fuel cell membranes, and electronics, tissue engineering processes [6, 8, 10]. Electrospinning has proven to be the best nanofiber manufacturing process because of simplicity and material compatibility [11]. We give a general outlook of mathematical models for electrospinning of conducting polymers in this chapter.

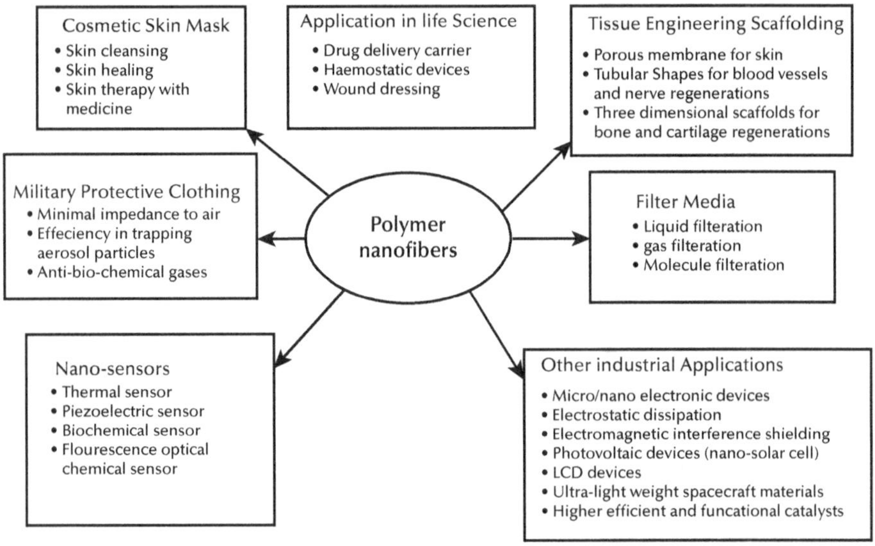

FIGURE 1 Potential application of electrospun polymer nanofibers.

Generally in this process (Figure 2), a polymer solution or melt is supplied through a syringe ~10–20 cm above a grounded substrate. The process is driven by an electrical potential of the order of kilovolts applied between the syringe and the substrate [10]. Electrospinning of polymer solutions involves to a first approximation, a rapid evaporation of the solvent. The evaporation of the solvent thus will happen on a time scale well below the second range [12]. The elongation of the jet during electrospinning is initiated by electrostatic force, gravity, inertia, viscosity, and surface tension [1]. Unlike, traditional spinning process which principally uses gravity and externally applied tension, electrospinning uses externally applied electric field, as driving force [5, 11].

The applied voltage induces a high electric charge on the surface of the solution drop at the needle tip. The drop now experiences two major electrostatic forces: The coulombic force which is induced by the electrical field, and electrostatic repulsion between the surface charges. The intensity of these forces causes the hemispherical surface of the drop to elongate and form a conical shape, which is also known as the Taylor cone. By further increasing the strength of the field, the electrostatic forces in the drop will overcome the surface tension of the fluid/air interface and an electrically charged jet will be ejected [13]. After the ejection, the jet elongates and the solvent evaporates, leaving an electrically charged fiber which during the elongation process becomes very thin [1, 2, 12, 14]. Study of effects of various electrospinning parameters is important to improve the rate of nanofiber processing. In addition, several applications demand well-oriented nanofibers [11]. Theoretical studies about the stability of an isolated charged liquid droplet predicted that it becomes unstable and fission takes place when the charge becomes sufficiently large compared to the stabilizing effect of the surface tension [15, 16].

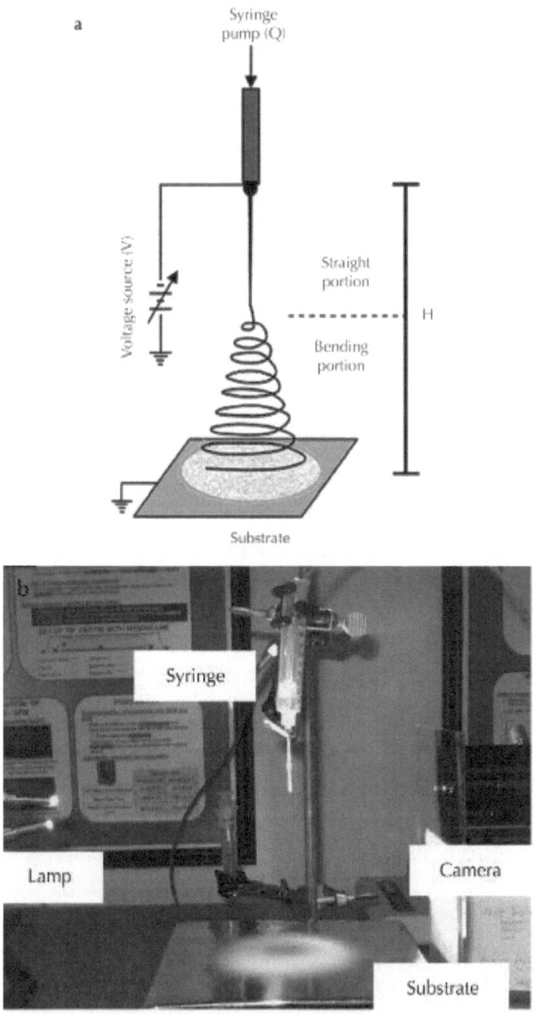

FIGURE 2 A scheme of electrospinning set up for nanofiber production.

The characteristic feature of this process is the onset of a chaotic oscillation of the elongating jet which is due to the electrostatic interactions between the external electric field and the surface charges on the jet as well as the electrostatic repulsion of mutual fiber parts. The fiber can be spun directly onto the grounded (conducting) screen or on an intermediate deposit material. Because of the oscillation, the fiber is deposited randomly on the collector, creating a so called "nonwoven" fiber fabric [1, 17]. An important stage of nanofibers formation in electrospinning include fluid instabilities such as whipping instabilities [17]. The applied voltage V, fluid flow rate Q, and separation distance H, are manipulated such that a steady, electrostatically driven jet of fluid is drawn from the capillary tip and is collected upon the grounded substrate [2, 10, 17]. These instabilities depend on fluid parameters and equipment configuration such as location of electrodes and the form of spinneret [17].

In melt or dry/wet solution spinning the shape and diameter of the die, as well as mechanical forces inducing specific draw ratios and drawing speeds highly determine dimensional and structural properties of the final fibers [18]. Properties that are known to significantly affect the electrospinning process are the polymer molecular weight, the molecular-weight distribution, the architecture (branched, linear, etc.) of the polymer, temperature, humidity, and air velocity in the chamber and processing parameters (like applied voltage, flow rate, types of collectors, tip to collector distance) as well as the rheological and electrical properties of the solution (viscosity, conductivity, surface tension, etc.) and finally motion of target screen [2, 6, 8, 19].

An important weak point of this method is a convective instability in the elongating jet. The jet will start rapidly whipping as it travels towards the collector. Therefore, during the electrospinning process, the whole substrate is covered with a layer of randomly placed fiber. The created fabric has a chaotic structure and it is difficult to characterize its properties [1, 11]. Electrospun fibers often have beads as "by products" [20]. Some polymer solutions are not readily electrospun when the polymer solution is too dilute due to limited solubility of the polymer. In these cases, the lack of elasticity of the solution prevents the formation of uniform fibers, instead, droplets or necklace like structures know as 'beads-on-string' are formed [21]. The electrospun beaded fibers are related to the instability of the jet of polymer solution. The bead diameter and spacing were related to: the fiber diameter, solution viscosity, net charge density carried by the electrospinning jet and surface tension of the solution [20, 21]. Important findings which were obtained during studies are:

(i) Fibers of different sizes, that is consisting of different numbers of parent chains, exhibit almost identical hyperbolic density profiles at the surfaces.
(ii) The end beads are predominant and the middle beads are depleted at the free surfaces.
(iii) There is anisotropy in the orientation of bonds and chains at the surface.
(iv) The centre of mass distribution of the chains exhibits oscillatory behavior across the fibers.
(v) The mobility of the chain in nanofiber increases as the diameter of the nanofiber decreases [19].

It is necessary to develop theoretical and numerical models of electrospinning because of demanding a different optimization procedure for each material [8]. Modeling and simulation of electrospinning process will help to understand the following:

(a) The cause for whipping instability.
(b) The dependence of jet formation and jet instability on the process parameters and fluid properties for better jet control and higher production rate.
(c) The effect of secondary external field on jet instability and fiber orientation [2, 11].

Several techniques such as dry rotary electrospinning [22] and scanned electrospinning nanofiber deposition system [23] control deposition of oriented nanofibers.

In the following, some basic and necessary theories for electrospinning modeling are reviewed.

16.2 BASICS OF ELECTROSPINNING MODELING

Modeling of the electrospinning process will be useful for the factors perception that cannot be measured experimentally [24]. Although electrospinning gives continuous fibers, mass production, and the ability to control nanofibers properties are not obtained yet. In electrospinning the nanofibers for a random state on the collector plate while in many applications of these fibers such as tissue engineering well-oriented nanofibers are needed. Modeling and simulations will give a better understanding of electrospinning jet mechanics [11]. The development of a relatively simple model of the process has been prevented by the lack of systematic, fully characterized experimental observations suitable to lead and test the theoretical development [17]. The governing parameters on electrospinning process which are investigated by modeling are solution volumetric flow rate, polymer weight concentration, molecular weight, the applied voltage, and the nozzle to ground distance [1, 7, 25]. The macroscopic nanofiber properties can be determined by multiscale modeling approach. For this purpose, at first the effective properties determined by using modified shear lag model then by using of volumetric homogenization approach, the macro scale properties concluded [26].

Till date two important modeling zones have been introduced. These zones are: (a) The zone close to the capillary (jet initiation zone) outlet of the jet and (b) The whipping instability zone where the jet spirals and accelerates towards the collector plate [11, 24, 27].

The parameters influence the nature and diameter of the final fiber so obtaining the ability to control them is a major challenge. For selected applications it is desirable to control not only the fiber diameter, but also the internal morphology [28]. An ideal operation would be: the nanofibers diameter to be controllable, the surface of the fibers to be intact and a single fiber would be collectable. The control of the fiber diameter can be affected by the solution concentration, the electric field strength, the feeding rate at the needle tip and the gap between the needle and the collecting screen (Figure 3) [1, 12, 29].

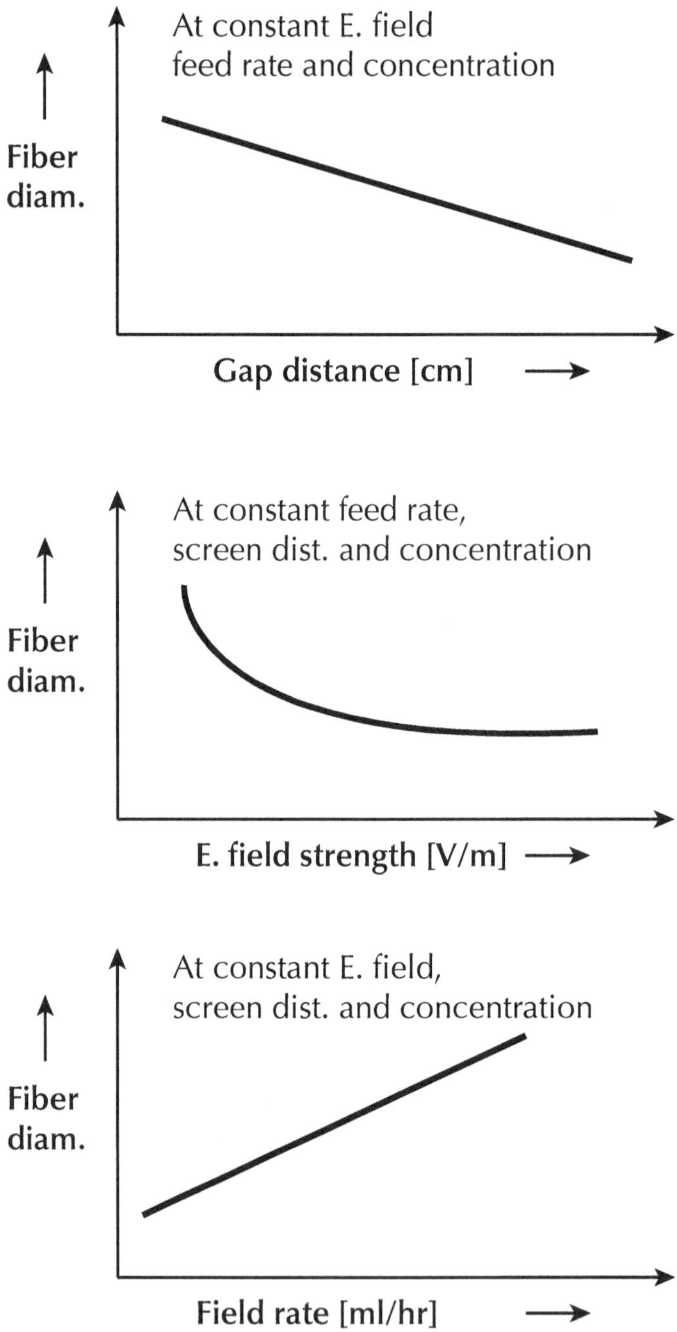

FIGURE 3 *(Continued)*

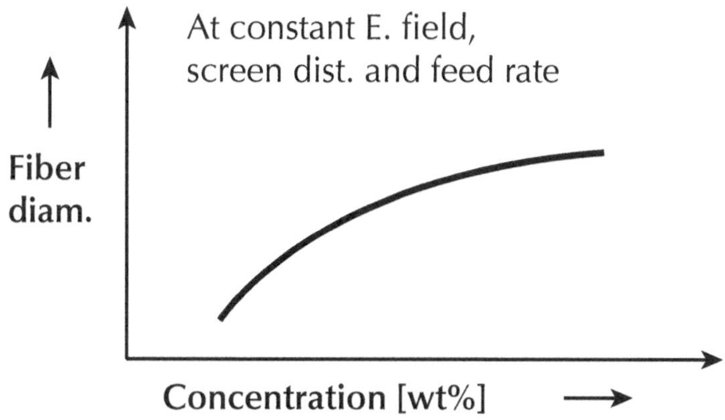

FIGURE 3 Effect of process parameters on fiber diameter.

Control over the fiber diameter remains a technological bottleneck. However A cubic model for mean fiber diameter was developed for samples by A. Doustgani et al. A suitable theoretical model of the electrospinning process is one that can show a strong-moderate-minor rating effects of these parameters on the fiber diameter. Some disadvantages of this method are low production rate, non oriented nanofiber production, difficulty in diameter prediction, and controlling nanofiber morphology, absence of enough information on rheological behavior of polymer solution and difficulty in precise process control that emphasis necessity of modeling [25, 29–30].

16.2.1 Viscoelastic Flows Analysis

Recently, significant progress has been made in the development of numerical algorithms for the stable and accurate solution of viscoelastic flow problems, which exits in processes like electrospinning process. A limitation is made to mixed finite element methods to solve viscoelastic flows using constitutive equations of the differential type [31].

Governing Set of Equations

The analysis of viscoelastic flows includes the solution of a coupled set of partial differential equations: The equations depicting the conservation of mass, momentum, energy, and constitutive equations for a number of physical quantities present in the conservation equations such as density, internal energy, heat flux, stress, and so on depend on process [31].

Approaches to Viscoelastic Finite Element Computations

There are fundamentally different approaches like: A mixed formulation may be adopted different parameters like velocity, pressure, and so on including the constitutive equation is multiplied independently with a weighting function and transformed in a weighted residual form [31]. The constitutive equation may be transformed into an ordinary differential equation (ODE). For transient problems this can, for instance, be achieved in a natural manner by adopting a Lagrangian formulation [32].

Time Dependent Flows

By introducing a selective implicit/explicit treatment of various parts of the equations, a certain separating at each time step of the set of equations may be obtained to improve computational efficiency. This suggests the possibility to apply devoted solvers to subproblems of each fractional time step [31].

16.2.2 Basics of Hydrodynamics

Due to the reason that nanofibers are made of polymeric solutions forming jets it is necessary to have a basic knowledge of hydrodynamics [33]. According to the effort of finding a fundamental description of fluid dynamics, the theory of continuity was implemented. The theory describes fluids as small elementary volumes which still are consisted of many elementary particles.

The equation of continuity:

$$\frac{\partial \rho_m}{\partial t} + div(\rho_m v) = 0 \quad \text{(For incompressible fluids } div(v) = 0\text{)}$$

The Euler's equation simplified for electrospinning:

$$\frac{\partial v}{\partial t} + \frac{1}{\rho_m} \nabla p = 0$$

The equation of capillary pressure:

$$P_c = \frac{\gamma \partial^2 \zeta}{\partial x^2}$$

The equation of surface tension:

$$\Delta P = \gamma \left(\frac{1}{R_X} + \frac{1}{R_Y} \right)$$ R_x and R_y are radii of curvatures

The equation of viscosity:

$$\tau_{ij} = \eta\left(\frac{\partial v_i}{\partial x_j} + \frac{\partial v_j}{\partial x_i}\right) \quad \text{(For incompressible fluids, } \tau_{i,j} = \text{Stress tensor)}$$

$$V = \frac{\eta}{\rho_m} \quad \text{(kinematic viscosity)}$$

16.2.3 Electrohydrodynamic (EHD) Theory

In 1966 Taylor discovered that finite conductivity enables electrical charge to accumulate at the drop interface, permitting a tangential electric stress to be generated. The tangential electric stress drags fluid into motion, and thereby generates hydrodynamic stress at the drop interface. The complex interaction between the electric and hydrodynamic stresses causes either oblate or prolate drop deformation and in some special cases keeps the drop from deforming [34–35].

Feng has used a general treatment of Taylor-Melcher for stable part of electrospinning jets by one dimensional equations for mass, charge, and momentum. In this model a cylindrical fluid element is used to show electrospinning jet kinematic measurements [10].

In Figure 4 the essential parameters are: radius R, velocity v_z, electric field, E_z, total path length L, interfacial tension γ, interfacial charge σ, tensile stress τ, volumetric flow rate Q, conductivity K, density ρ, dielectric constant ε, and zero-shear rate viscosity η_0. The most important equation that Feng used are [10]:

$$\tilde{R}^2\tilde{\nu}_z = 1$$

$$\tilde{R}^2\tilde{E}_z + Pe_e\tilde{R}\tilde{\nu}_z\tilde{\sigma} = 1$$

$$\tilde{\upsilon}_z\tilde{\upsilon}'_z = \frac{1}{Fr} + \frac{\tilde{T}'}{\mathrm{Re}_j\tilde{R}^2} + \frac{1}{We}\frac{\tilde{R}'}{\tilde{R}^2} + \varepsilon(\tilde{\sigma}\tilde{\sigma}' + \beta\tilde{E}_z\tilde{E}'_z + \frac{2\tilde{\sigma}\tilde{E}_z}{\tilde{R}})$$

$$\tilde{E}_z = \tilde{E}_0 - \ln\chi\left[(\tilde{\sigma}\tilde{R})' - \frac{\beta}{2}(\tilde{E}\tilde{R}^2)''\right]$$

$$E_0 = {}^{\eta_0\nu_0}\!/_{R_0} \qquad \beta - ({}^{\varepsilon}\!/_{\bar{\varepsilon}}) - 1 \qquad \tau = R^2(\tilde{\tau}_{zz} - \tilde{\tau}_{rr})$$

Feng solved equation under different fluid properties, particularly for non-Newtonian fluids with extensional thinning, thickening, and strain hardening but Helgeson et al. developed a simplified understanding of electrospinning jets based on the evolution of the tensile stress due to elongation [10].

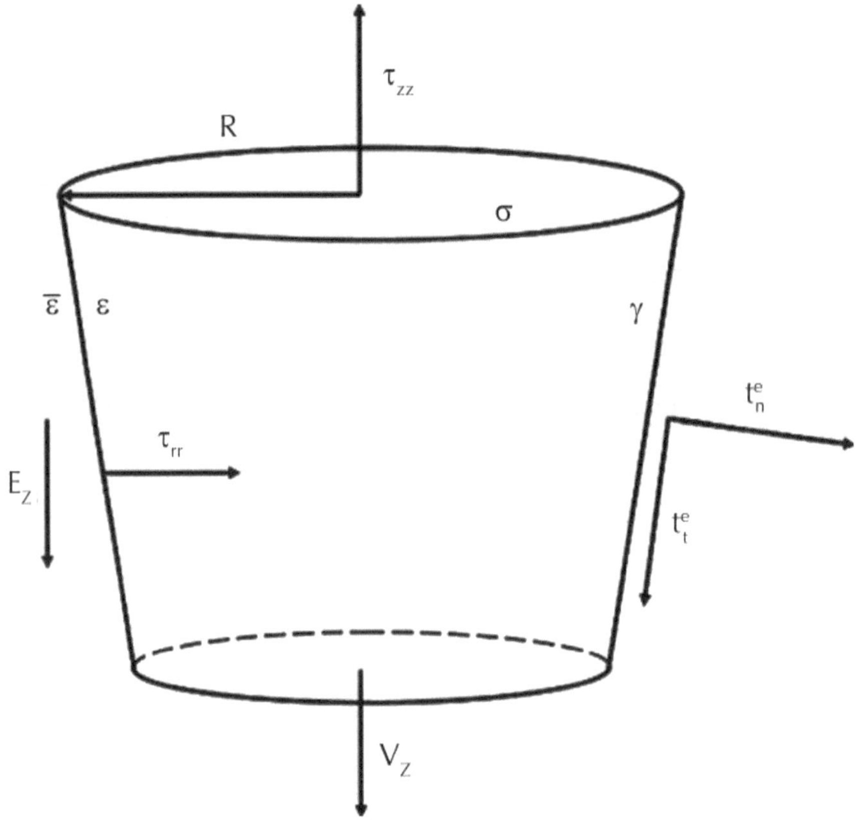

FIGURE 4 Scheme of the cylindrical fluid element used in EHD modeling.

16.2.4 Electric Forces in Fluids

The initialization of instability on the surfaces of liquids should be caused using the application of the external electric field that induces electric forces on surfaces of liquids. A localized approximation was developed to calculate the bending electric force acting on an electrified polymer jet which is an important element of the electrospinning process for manufacturing of nanofibers. Using this force, a far reaching analogy between the electrically driven bending instability and the aerodynamically driven instability was established. The description of the wave's instabilities is expressed by equations called dispersion laws. The dependence of wavelength on the surface tension γ is almost linear and the wavelengths between jets are a little bit smaller for lower depths. Dependency of wavelength on electric field strength is exponential. The dispersion law is identified for four groups of dielectrics liquids with using of Clausius-Mossotti and Onsager's relation (nonpolar liquids with finite and infinite depth and weakly polar liquids with finite and infinite depth). According to these relations relative permittivity is a function of parameters like temperature, square of angular frequency, wave length number, and reflective index [2, 33, 36, 37].

16.2.5 Dimensionless Non-Newtonian Fluid Mechanics

The best way for analyzing fluid mechanics problems is converting parameters to dimensionless form. By using this method the numbers of governing parameters for given geometry and initial condition reduce. The nondimensionalization of a fluid mechanics problem generally starts with the selection of a characteristic velocity then because the flow of non-Newtonian fluids the stress depends non-linearly on the flow kinematics, the deformation rate is a main quantity in the analysis of these flows. Next step after determining different parameters is evaluate characteristic values of the parameters. Then the nondimensionalization procedure is to employ the selected characteristic quantities to obtain a dimensionless version of the conservation equations and get to certain numbers like Reynolds number and the Galilei number. The excessive number of governing dimensionless groups poses certain difficulties in the fluid mechanics analysis. Finally by using these results in equation and applying boundary conditions it can be achieved to study different properties [38].

16.2.6 Detection of X-ray Generated by Electrospinning

Electrospinning jets producing nanofibers from a polymer solution by electrical forces are fine cylindrical electrodes that create extremely high electric-field strength in their vicinity at atmospheric conditions. However, this quality of electrospinning is only scarcely investigated, and the interactions of the electric fields generated by them with ambient gases are nearly unknown. Pokorny et al. reported on the discovery that electrospinning jets generate X-ray beams up to energies of 20 keV at atmospheric conditions. [39]. Mikes et al. investigated on the discovery that electrically charged polymeric jets and highly charged gold-coated nanofibrous layers in contact with ambient atmospheric air generate X-ray beams up to energies of hard X-rays[40–41].

The first detection of X-ray produced by nanofiber deposition was observed using radiographic films. The main goal of using these films understands of Taylor cone creation. The X-ray generation is probably dependent on diameters of the nanofibers that are affected by the flow rate and viscosity. So it is important to find the ideal concentration (viscosity) of polymeric solutions and flow rate to spin nanofibers as thin as possible. The X-ray radiation can produce black traces on the radiographic film. These black traces had been made by outer radiation generated by nanofibers and the radiation has to be in the X-ray region of electromagnetic spectra, because the radiation of lower energy is absorbed by the shield film. Radiographic method of X-ray detection is efficient and sensitive. It is obvious that this method didn't tell us anything about its spectrum, but it can clearly show its space distribution. The humidity, temperature and rheological parameters of polymer can affect on the X-ray intensity generated by nanofibers [33]. The necessity of modeling in electrospinning process and a quick outlook of some important models will be discussed as follows.

16.3 MODELING ELECTROSPINNING OF NANOFIBERS

Using theoretical prediction of the parameter effects on jet radius and morphology can significantly reduce experimental time by identifying the most likely values that will yield specific qualities prior to production [25]. All models start with some assumptions and have short comings that need to be addressed [14]. The basic principles for dealing with electrified fluids that Taylor discovered, is impossible to account for most electrical phenomena involving moving fluids under the seemingly reasonable assumptions that the fluid is either a perfect dielectric or a perfect conductor. The reason is that any perfect dielectric still contains a nonzero free charge density. The presence of both an axisymmetric instability and an oscillatory "whipping" instability of the centerline of the jet; however, the quantitative characteristics of these instabilities disagree strongly with experiments [19, 36]. During steady jetting, a straight part of the jet occurs next to the Taylor cone, where only axisymmetric motion of the jet is observed. This region of the jet remains stable in time. However, further along its path the jet can be unstable by non-axisymmetric instabilities such as bending and branching, where lateral motion of the jet is observed in the part near the collector [10].

Branching as the instability of the electrospinning jet can happen quite regularly along the jet if the electrospinning conditions are selected appropriately. Branching is a direct consequence of the presence of surface charges on the jet surface, as well as of the externally applied electric field. The bending instability leads to drastic stretching and thinning of polymer jets towards nanoscale in cross section. Electrospun jets also caused to shape perturbations similar to undulations which can be the source of various secondary instabilities leading to nonlinear morphologies developing on the jets [18]. The bending instabilities that occur during electrospinning have been studied and mathematically modelled by Reneker et al. by viscoelastic dumbbells connected together [42]. Both electrostatic and fluid dynamic instabilities can contribute to the basic operation of the process [19].

Different stages of electrospun jets have been investigated by different mathematical models during last decade by one or three dimensional techniques [8, 43].

Physical models which study the jet profile, the stability of the jet and the cone like surface of the jet have been develop due to significant effects of jet shape on fiber qualities [7]. Droplet deformation, jet initiation and in particular, the bending instability which control to a major extent fiber sizes and properties are controlled apparently predominantly by charges located within the flight jet [18]. An accurate, predictive tool using a verifiable model that accounts for multiple factors would provide a means to run many different scenarios quickly without the cost and time of experimental trial and error [25].

16.3.1 An Outlook to Significant Models

The models typically treat the jet mechanics using the localized induction approximation by analogy to aerodynamically driven jets. They include the viscoelasticity of the spinning fluid and have also been augmented to account for solvent evaporation in the jet. These models can describe the bending instability and fiber morphology. Because of difficulty in measure model variables they cannot accurately design and control the electrospinning process [10]. Here is some current and more discussed models:

Leaky Dielectric Model

The principles for dealing with electrified fluids were summarized by Melcher and Taylor [36]. Their research showed that it is impossible to explain the most of the electrical phenomena involving moving fluids given the hypothesis that the fluid is either a perfect dielectric or a perfect conductor, since both the permittivity and the conductivity affect the flow. An electrified liquid always includes free charge. Although the charge density may be small enough to ignore bulk conduction effects, the charge will accumulate at the interfaces between fluids. The presence of this interfacial charge will result in an additional interfacial stress, especially a tangential stress which in turn will modify the fluid dynamics [36, 44].

The EHD theory proposed by Taylor as the leaky dielectric model is capable of predicting the drop deformation in qualitative agreement with the experimental observations [34, 35].

Although Taylor's leaky dielectric theory provides a good qualitative description for the deformation of a Newtonian drop in an electric field, the validity of its analytical results is strictly limited to the drop experiencing small deformation in an infinitely extended domain. Extensive experiments showed a serious difference in this theoretical prediction [34].

Some investigations have been done to solve this defect. For example, to examine electro kinetic effects, the leaky dielectric model was modified by consideration the charge transport [44, 45]. When the conductivity is finite, the leaky dielectric model can be used [45]. By use of this mean, Saville indicated that the solution is weakly conductive so the jet carries electric charges only on its surface [44, 45].

A Model for Shape Evaluation of Electrospinning Droplets

Comprehension of the drops behavior in an electric field is playing a critical role in practical applications. The electric field-driven flow is of practical importance in the processes in which improvement of the rate of mass or heat transfer between the drops and their surrounding fluid [34]. Numerically investigations about the shape evolution of small droplets attached to a conducting surface depended on strong electric fields (weak, strong, and super electrical) have done and indicated that three different scenarios of droplet shape evolution are distinguished, based on numerical solution of the Stokes equations for perfectly conducting droplets by investigation of Maxwell stresses and surface tension [13, 46]. The advantages of this model are that the non-Newtonian effect on the drop dynamics is successfully identified on the basis of electrohydrostatics at least qualitatively. In addition, the model showed that the deformation and breakup modes of the non-Newtonian drops are distinctively different from the Newtonian cases [34].

Nonlinear Model

A simple two-dimensional model can be used for description of formation of barb electrospun polymer nanowires with a perturbed swollen cross-section and the electric charges "frozen" into the jet surface. This model was integrated numerically using the Kutta-Merson method with the adoptable time step. The result of this modeling is explained theoretically as a result of relatively slow charge relaxation compared to the development of the secondary electrically driven instabilities which deform jet surface locally. When the disparity of the slow charge relaxation compared to the rate of growth of the secondary electrically driven instabilities becomes even more pronounced, the barbs transform in full scale long branches. The competition between charge relaxation and rate of growth of capillary and electrically driven secondary localized perturbations of the jet surface is affected not only by the electric conductivity of polymer solutions but also by their viscoelasticity. Moreover, a nonlinear theoretical model was able to resemble the main morphological trends recorded in the experiments [18].

A Mathematical Model for Electrospinning Process under Coupled Field Forces

There is not a theoretical model which can describe the electrospinning process under the multi field forces so a simple model might be very useful to indicate the contributing factors. Modeling this process can be done in two ways:

(a) The deterministic approach which uses classical mechanics like Euler approach and Lagrange approach.
(b) The probabilistic approach uses E-infinite theory and quantum like properties.

Many basic properties are harmonious by adjusting electrospinning parameters such as voltage, flow rate and others, and it can offer in-depth inside into physical understanding of many complex phenomena which cannot be fully explain [9].

Slender-body Model

One-dimensional models for inviscid, incompressible, axisymmetric, annular liquid jets falling under gravity have been obtained by means of methods of regular perturbations for slender or long jets, integral formulations, Taylor's series expansions, weighted residuals, and variational principles [27, 47].

Using Feng's theory some familiar assumptions in modeling jets and drops are applied: The jet radius R decreases slowly along z direction while the velocity v is uniform in the cross section of the jet so it is lead to the nonuniform elongation of jet. According to the parameters can be arranged into three categories: process parameters (Q, I, and E_∞), geometric parameters (R_0 and L) and material parameters (ρ, η_0 (the zero-shear-rate viscosity), ε, ε, K, and γ). The jet can be represented by four steady state equations: the conservation of mass and electric charges, the linear momentum balance and Coulomb's law for the E field.

Mass conservation can be stated by:

$$\pi R^2 v = Q$$

R: Jet radius

The second equation in this modeling is charge conservation that can be stated by:

$$\pi R^2 KE + 2\pi R \nu \sigma = I$$

The linear momentum balance is:

$$\rho v v' = \rho g + \frac{3}{R^2}\frac{d}{dz}(\eta R^2 v) + \frac{\gamma R'}{R^2} + \frac{\sigma \sigma'}{\bar{\varepsilon}} + (\varepsilon - \bar{\varepsilon})EE' + \frac{2\sigma E}{R}$$

The Coulomb's law for electric field:

$$E(z) = E_{\infty}(z) - \ln \chi(\frac{1}{\bar{\varepsilon}}\frac{d(\sigma R)}{dz} - \frac{\beta}{2}\frac{d^2(ER^2)}{dz^2})$$

L: The length of the gap between the nozzle and deposition point
R_0: The initial jet radius

$$\beta = \varepsilon / \bar{\varepsilon} - 1$$

$$\chi = L / R_0$$

By these four equations the four unknown functions R, v, E, and σ are identified.

At first step the characteristic scales such as R_0 and v_0 are denoted to format dimensionless groups. Inserting these dimensionless groups in four equations discussed the final dimensionless equations is obtained. The boundary conditions of four equations which became dimensionless can be expressed (see Figure 5) as:

$$\ln z = 0 \qquad R(0) = 1 \qquad \sigma(0) = 0$$

$$\ln z = \chi \qquad E(\chi) = E_{\infty}$$

The first step is to write the ODE's as a system of first order ODE's by a numerical relaxation method as for example solve from numerical recipes. The basic idea is to introduce new variables, one for each variable in the original problem and one for each of its derivatives up to one less than the highest derivative appearing. For solving this ODE the Fortran program is used, in the first step an initial guess uses for χ and the other parameters would change according to χ [1, 48]. The limitation of slender body theory is: avoiding treating physics near the nozzle directly [27].

A Model for Electrospinning Viscoelastic Fluids

When the jet thins, the surface charge density varies, which in turn affects the electric field and the pulling force. Roozemond combined the "leaky dielectric model" and the "slender-body approximation" for modeling electrospinning viscoelastic jet [41]. The jet could be represented by four steady state equations: the conservation of mass and electric charges, the linear momentum balance and Coulomb's law for the electric field, with all quantities depending only on the axial position z. The equations can be converted to dimensionless form using some characteristic scales and dimensionless groups like most of the models. These equations could be solved by converting to ODE's forms and using suitable boundary conditions [27, 48].

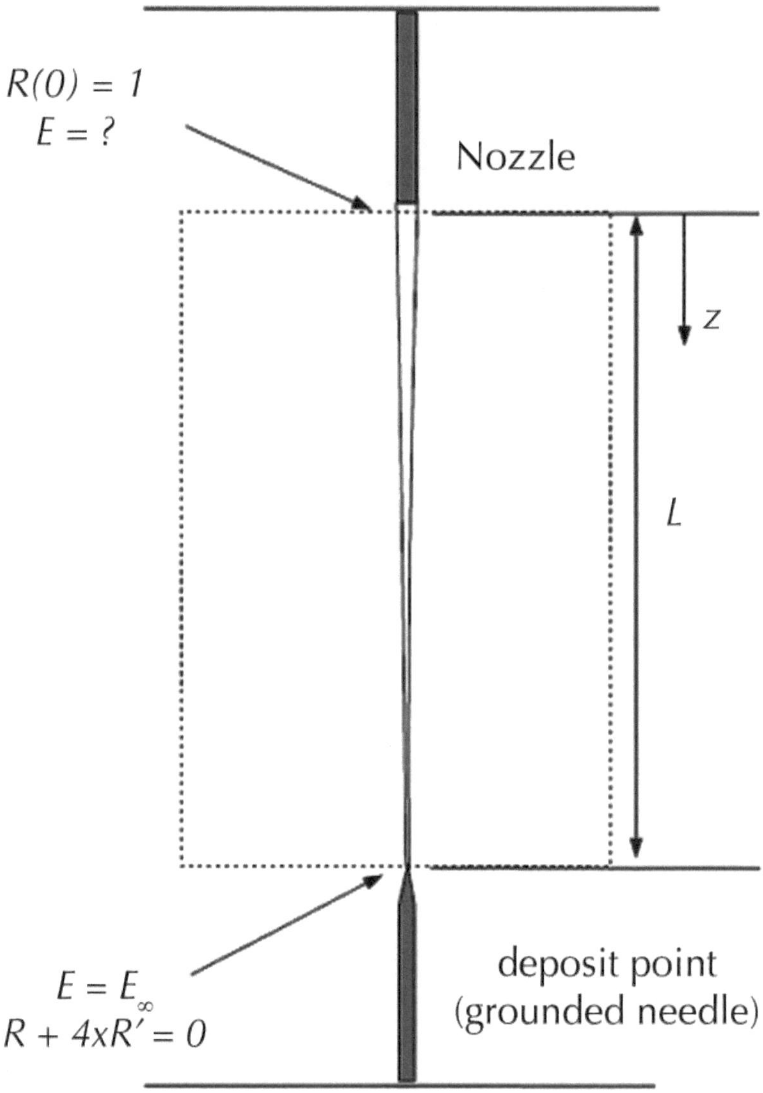

FIGURE 5 Boundary conditions.

Lattice Boltzmann Method (LBM)

Developing LBM instead of traditional numerical techniques like finite volume, finite difference, and finite element for solving large scale computations and problems involving complex fluids, colloidal suspensions, and moving boundaries is so useful [11].

Mathematical Model for AC-electrospinning

Much of the nanofiber research reported so far was on nanofibers made from DC potential [7]. In DC-electrospinning, the fiber instability or 'whipping' has made it difficult to control the fiber location and the resulting microstructure of electrospun materials. To overcome these limitations, some new technologies were applied in the electrospinning process. The investigations proved that AC potential resulted in a significant reduction in the amount of fiber 'whipping' and the resulting mats exhibited a higher degree of fiber alignment but were observed to contain more residual solvent. In AC-electrospinning, the jet is inherently unsteady due to the AC potential unlike DC ones, so all thermal, electrical, and hydrodynamics parameters was considered to be effective in the process [49, 50].

The governing equations for an unsteady flow of an infinite viscous jet pulled from a capillary orifice and accelerated by an AC potential can be expressed as follows:

(1) The conservation of mass equation
(2) Conservation of charge
(3) The Navier-Stokes equation

Using these governing equations, final model of AC electrospinning is able to find the relationship between the radius of the jet and the axial distance from nozzle and a scaling relation between fiber radius and the AC frequency[50].

Multiple Jet Modeling

It was experimentally and numerically exhibited that the jets from multiple nozzles expose higher repulsion by another jets from the neighborhood by columbic forces than jets spun by a single nozzle process [5]. Yarin and Zussman achieved upward electrospinning of fibers from multiple jets without the use of nozzles, instead using the spiking effect of a magnetic liquid [51]. For large scale nanofiber production and the increase in production rate, multi jet electrospinning systems can be designed to increase both productivity and covering area [52, 53]. The linear Maxwell model and nonlinear Upper-convected Maxwell (UCM) model were used to calculate the viscoelasticity. By using these models the primary and secondary bending instabilities can be calculated. Maxwell model and the nonlinear UCM model lead to rather close results in the flow dominated by the electric forces. In a multiple nozzle set up, not only the external applied electric field and self induced coulombic interactions influence the jet path, but also mutual coulombic interactions between different jets contribute [53].

A Mathematical Model of the Magnetic Electrospinning Process

For controlling the instability, magnetic electrospinning is proposed. For describing the magnetic electrospun jet, it can be used Reneker's model [42]. This model does not consider the coupling effects of the thermal field, electric field, and magnetic field. Therefore, the momentum equation for the motion of the beads is [54]:

$$m\frac{d^2 r_i}{dt^2} = F_C + F_E + F_{ve} + F_B + F_q$$

16.4 CONCLUSION

Electrospinning is a very simple and versatile method of creating polymer based high functional and high performance nanofibers that can revolutionize the world of structural materials. The process is versatile in that there is a wide range of polymer and biopolymer solutions or melts that can spin. The electrospinning process is a fluid dynamics related problem. In order to control the property, geometry, and mass production of the nanofibers, it is necessary to understand quantitatively how the electrospinning process transforms the fluid solution through a millimeter diameter capillary tube into solid fibers which are four to five orders smaller in diameter. When the applied electrostatic forces overcome the fluid surface tension, the electrified fluid forms a jet out of the capillary tip towards a grounded collecting screen. Although electrospinning gives continuous nanofibers, mass production, and the ability to control nanofibers properties are not obtained yet. Combination of both theoretical and experimental approaches seems to be promising step for better description of electrospinning process. Applying simple models of the process can be useful in atoning the lack of systematic, fully characterized experimental observations, and the theoretical aspects in predicting and controlling effective parameters. The analysis and comparison of model with experiments identify the critical role of the spinning fluid's parameters. The theoretical and quantitative tools developed in different models provide semi empirical methods for predicting ideal electrospinning process or electrospun fiber properties. In each model, researcher tried to improve the existing models or changed the tools in electrospinning by using another view. Therefore, we attempted to have a whole view on important models after investigation about basic objects. A real mathematical model or more accurately, a real physical model, might initiate a revolution in understanding of dynamic and quantum-like phenomena in the electrospinning process. A new theory is much needed which bridges the gap between Newton's world and the quantum world.

KEYWORDS

- **Dispersion law**
- **Electrohydrodynamic**
- **Electrospinning**
- **Lattice Boltzmann method**
- **Leaky dielectric model**
- **Navier-Stokes equation**

REFERENCES

1. Solberg, R. H. M. *Position-controlled deposition for electrospinning in Department Mechanical Engineering*. Eindhoven University of Technology, Eindhoven p. 67 (2007).
2. Chronakis, I. S. Processing, Properties and Applications, in Micro-/Nano-Fibers by Electrospinning Technology. pp. 264–286.

3. Formhals, A. *Process and apparatus for preparing artificial threads*. U.S. Patent, Editor. Germany (1934).
4. Reneker, D. H. and Chun, I. Nanometer diameter fibers of polymer produced by electrospinning. *Nanotechnology*, **7**, 216–223 (1996).
5. Kim, G., Cho, Y. S., and Kim, W. D. Macromolecular Nanotechnology, Stability analysis for multi-jets electrospinning process modified with a cylindrical electrode. *European Polymer Journal*, **42**, 2031–2038 (2006).
6. Bhardwaj, N. and Kundu, S. C. Research review paper, Electrospinning: A fascinating fiber fabrication technique. *Biotechnology Advances*, **28**, 325–347 (2010).
7. Theron, S. A., Zussman, E., and Yarin, A. L. Experimental investigation of the governing parameters in the electrospining of polymer solutions. *Polymer*, **45**, 2017–2030 (2004).
8. Kowalewski, T. A., Barral, S., and Kowalczyk, T. Modeling Electrospinning of Nanofibers. *IUTAM Symposium on Modelling Nanomaterials and Nanosystems*, **13**, 279–293 (2009).
9. Xu, L. A mathematical model for electrospinning process under coupled field forces. *Chaos, Solitons and Fractals*, **42**, 1463–1465 (2009).
10. Helgeson, E. M., et al. Theory and kinematic measurements of the mechanics of stable electrospun polymer jets. *Polymer*, **49**, 2924–2936 (2008).
11. Karra, S. Modeling electrospinning process and a numerical scheme using lattice Boltzmann method to simulate viscoelastic fluid flows, in Mechanical Engineering. Indian Institute of Technology Madras, Chennai. p. 60 (2007).
12. Bognitzki, M. et al. Nanostructured Fibers via Electrospinning. *Advanced Materials*, **13**, 70–73 (2001).
13. Basaran, O. A. and Suryo, R. Fluid Dynamics the invisible jet. *Nature Physics*, **3**, 679–680 (2007).
14. Titchenal, N. and Schrepple, W. Modeling of Electro-spinning. *Materials Science and Engineering*.
15. Yarin, A. L., Koombhongse, S., and Reneker, D. H. Taylor cone and jetting from liquid droplets in electrospinning of nanofibers. *Journal of Applied Physics*, **90**, 4836–4846 (2001).
16. Papageorgous, D. T. and Petropoulos, P. G. Generation of interfacial instabilities in charged electrified viscous liquid films. *Journal of Engineering Mathematics*, **50**, 223–240 (2004).
17. Shin, Y. M. et al. Experimental characterization of electrospinning the electrically forced jet and instabilities. *Polymer*, **42**, 9955–9967 (2001).
18. Holzmeister, A., Yarin, A. L., and Wendorff, J. H. Barb formation in electrospinning Experimental and theoretical investigations. *Polymer*, **51**, 2769–2778 (2010).
19. Frenot, A. and Chronakis, I. S. Polymer nanofibers assembled by electrospinning. *Current Opinion in Colloid and Interface Science*, **8**, 64–75 (2003).
20. Fong, H., Chun, I., and Reneker, D. H. Beaded nanofibers formed during electrospinning. *Polymer*, **40**, 4585–4592 (1999).
21. Yu, H. J., Fridrikh, S. V., and Rutledge, G. C. The role of elasticity in the formation of electrospun fibers. *Polymer*, **47**, 4789–4797 (2006).
22. El-Auf, A. K. Nanofibers and nanocomposites poly (3,4-ethylene dioxythiophene)/poly(styrene sulfonate) by electrospinning, in Department of Materials Science and Engineering. Drexel University, Philadelphia p. 261 (2004).
23. Czaplewski, D., Kameoka, J., and Craighead, H. G. Nonlithographic approach to nanostructure fabrication using a scanned electrospinning source. *Journal of Vacuum Science & Technology B (Microelectronics and Nanometer Structures)*, **21**, 2994–2997 (2003).
24. Patanaik, A., Jacobs, V., and Anandjiwala, R. D. Experimental study and modeling of the electrospinning process. In *86th Textile Institute World Conference*. Hong Kong, pp. 1160–1168 (2008).
25. Thompson, C. J. et al. Effects of parameters on nanofiber diameter determined from electrospinning model. *Polymer*, **48**, 6913–6922 (2007).
26. Agic', A. Multiscale Modeling electrospun nanofiber structure. *Materials Science Forum*, **714**, 33–40 (2012).

27. Feng, J. J. The stretching of an electrified non-Newtonian jet a model for electrospinning. *Physics of Fluid*, **14**, 3912–3926 (2002).
28. Helgeson, M. E. and Wagner, N. J. *A Correlation for the Diameter of Electrospun Polymer Nanofibers*. American Institute of Chemical Engineers, **53**, 51–55 (2007).
29. Doustgani, A., et al. Optimizing the mechanical properties of electrospun polycaprolactone and nanohydroxyapatite composite nanofibers. *Composites Part B*, **43**, 1830–1836 (2012).
30. Fridrikh, S. V., et al. Controlling the Fiber Diameter during Electrospinning. *Physical Review Letters*, **90**, 144502–4 (2003).
31. Baaijens, F. P. T. *Mixed finite element methods for viscoelastic flow analysis: A review, in Faculty of Mechanical Engineering*. Eindhoven University of Technology Center for Polymers and Composites, Eindhoven p. 37 (2001).
32. Rasmussen, H. K. and Hassager, O. Simulation of transient viscoelastic flow. *Journal of Non-Newtonian Fluid Mechanic*, **46**, 298–305 (1993).
33. Mikeš, I. P. *Physical principles of electrostatic spinning, in Physical engineering*. Technical University in Liberec, Liberec p. 122 (2011).
34. Ha, J. W. and Yang, S. M. Deformation and breakup of Newtonian and non-Newtonian conducting drops in an electric field. *Journal of Fluid Mechanics*, **405**, 131–156 (2000).
35. Taylor, G. I. Studies in electrohydrodynamics. I. The circulation produced in a drop by an electric field. Proceedings of the Royal Society of London. Series A. *Mathematical and Physical Sciences*, **291**, 159–166 (1966).
36. Melcher, J. R. and Taylor, G. I. Electrohydrodynamics A review of the role of interfacial shear stresses. *Annual Review of Fluid Mechanics*, **1**, 111–146 (1969).
37. Yarin, A. L., Koombhongse, S., and Reneker, D. H. Bending instability in electrospinning of nanofibers. *Journal of Applied Physics*, **89**, 3018–3026 (2001).
38. De Souza Mendes, R. P. Dimensionless non-Newtonian fluid mechanics. *Journal Non-Newtonian Fluid Mechanics*, **147**, 109–116 (2007).
39. Pokorn´y, P., Mikes, P., and Luk´aˇs, D. Electrospinning jets as X-ray sources at atmospheric conditions. *A Letter Journal Exploring the Frontiers of Physics*, **92**, 47002–47007 (2010).
40. MIikes, P., et al. High Energy Radiation Emitted from Nanofibers, in 7th International Conference - TEXSCI 2010. Czech Republic, Liberec (2010).
41. Kornev, K. G. Electrospinning Distribution of charges in liquid jets. *Journal of Applied Physics*, **110**, 124910–124915 (2011).
42. Reneker, D. H. et al. Bending instability of electrically charged liquid jets of polymer solutions in electrospinning. *Journal of Applied Physics*, **87**, 4531–4547 (2000).
43. He, J. H., et al. Review, Mathematical models for continuous electrospun nanofibers and electrospun nanoporous microspheres. *Polymer International*, **56**, 1323–1329 (2007).
44. Saville, D. A. Electrohydrodynamics the Taylor-Melcher leaky dielectric model. *Annual Review of Fluid Mechanics*, **29**, 27–64 (1997).
45. Parageorgiou, T. D. and Broeck, J. M. V. Numerical and Analytical Studies of Non-linear Gravity-Capillary Waves in Fluid Layers Under Normal Electric Fields. *IMA Journal of Applied Mathematics*, **72**, 832–853 (2007).
46. Reznik, S. N., et al. Transient and steady shapes of droplets attached to a surface in a strong electric field. *Journal of Fluid Mechanics*, **516**, 349–377 (2004).
47. Ramos, J. I. One-dimensional models of steady, inviscid, annular liquid jets. *Applied Mathematical Modelling*, **20**, 593–607 (1996).
48. Roozemond, P. C. *A Model for Electrospinning Viscoelastic Fluids*, in Department of Mechanical Engineering. Eindhoven University of Technology, Eindhoven p. 26 (2007).
49. Shin, Y. M., et al. Electrospinning A whipping fluid jet generates submicron polymer fibers. *Applied Physics Letters*, **78**, 3–7 (2001).
50. Ji-Huan, H., Yue, W., and Ning, P. A Mathematical Model for Preparation by AC-Electrospinning Process. *International Journal of Nonlinear Sciences and Numerical Simulation*, **6**, 243–248 (2005).

51. Yarin, A. L. and Zussman, E. Upward electrospinning of multiple nanofibers without needles/ nozzles. *Polymer*, **45**, 2977–2980 (2004).
52. Varesano, A., Carletto, R. A., and Mazzuchetti, G. Experimental investigations on the multi-jet electrospinning process. *Journal of Materials Processing Technology*, **209**, 5178–5185 (2009).
53. Theron, S. A., et al. Multiple jets in electrospinning: experiment and modeling. *Polymer*, **46**, 2889–2899 (2005).
54. Xu, L., Wu, Y., and Nawaz, Y. Numerical study of magnetic electrospinning processes. *Computers and Mathematics with Applications*, **61**, 2116–2119 (2011).

Index

D

E

F

G

H

R